HQ
728
.E47

College of St. Scholastica, Depar...
...dy of Victory Hall, Duluth 5, ...

SO-BIP-411

WITHDRAWN
LIBRARY
COLLEGE OF ST. SCHOLASTICA
DULUTH 11, MINN.

NEUROTIC INTERACTION IN MARRIAGE

Editorial Advisory Board

———•••———

PAUL HOCH, M.D.
Commissioner of Mental Hygiene, State of New York

SOL W. GINSBURG, M.D.
Clinical Professor of Psychiatry, Albert Einstein Medical School, New York

LEOPOLD BELLAK, M.D.
Clinical Assistant Professor of Psychiatry, New York Medical College

LOUIS LINN, M.D.
Associate Attending Psychiatrist, Mount Sinai Hospital, New York

NEUROTIC INTERACTION IN MARRIAGE

EDITED BY

VICTOR W. EISENSTEIN, M.D.

PUBLISHERS   Basic Books, Inc.   NEW YORK

HQ
728
.E47

FIRST EDITION

COPYRIGHT, © 1956, BY BASIC BOOKS, INC.

ALL RIGHTS RESERVED

MANUFACTURED IN THE UNITED STATES OF AMERICA

DESIGNED BY SIDNEY SOLOMON

LIBRARY OF CONGRESS

CATALOG CARD NUMBER: 56–11606

B
—
B

Foreword

THIS VOLUME is devoted to the intensive study and investigation of the most fundamental unit of our social structure— the family.

Twenty-five recognized experts in the field of human relations have contributed their various pertinent approaches to the understanding and solution of marital maladjustment and its inevitable effect on family development. The authors have followed sound scientific principles, as far as this type of material will allow, in their discussions of the psychodynamics of marital relations and in dealing with the relation of cultural factors to marriage and family life. Their approach is straightforward and practical in terms of our present day psychological thinking, much of which is tentative and due for modification.

The over-all result is a volume which provides an unusually rich source of information, analytic aid and therapeutic guidance for workers who are involved in the study and attempted correction of the numerous problems of family life. *Neurotic Interaction in Marriage* represents a great deal of thought and investigation on the part of the able scholars who have collaborated to afford us an interpretation of one of our major social complications.

Nolan D. C. Lewis, M.D.
DIRECTOR OF RESEARCH,
NEW JERSEY NEUROPSYCHIATRIC INSTITUTE, PRINCETON

48643   LIBRARY
COLLEGE OF ST. SCHOLASTICA
DULUTH 11. MINN

Introduction

————◆◆◆————

THE CONCEPT of neurotic interaction is aptly illustrated in Schopenhauer's fable of the freezing porcupines who huddled together for warmth, but were repelled by the sting of each other's quills. Each time the need for warmth brought them together, their mutual irritation began anew. And so the porcupines were continually being driven together and forced apart because of their physical needs.

Similarly, the emotional reactions of human beings are intensified through intimacy and reciprocal influence so that the intervolving of marriage tends to complicate rather than solve the emotional problems of the individual. Nevertheless, neurotic people can and do make good marriages, while many relatively healthy persons contract discordant and unhappy unions. The course and outcome of a marriage are determined not merely by the personality difficulties of each partner but by the way the two personalities interact.

Statistically, there are some 400,000 divorces and annulments granted in this country each year, about forty per cent of which involve children. The official figures, however, do not begin to convey the true picture. We would have to add a far greater number of separations and desertions in order to appreciate the widespread consequences of broken homes in our society, where daily the courts, social agencies, mental hospitals and penal institutions bear witness to the ultimate effects of disturbed marital interaction.

In the pages of this book, the specialists who have studied the intricacies and specific patterns of interaction reveal the hidden sources of marital tension, and demonstrate the dovetailing as well as the clashing of unconscious emotional needs. Such a study necessarily transcends individual psychology and becomes, in a sense, a social psychology which explores the dynamics of the marital partnership and its effect upon the family unit.

For the most part, these contents are based on investigation sponsored by the Psychiatric Forum Group of New York in the years 1949–1955, during which outstanding psychoanalysts, clinical psychiatrists, psychologists and psychiatric social workers exchanged views based on years of experience in their particular professions. Their contributions from private practice and agency experience represent a wealth of first-hand observations on people in all walks of life. Additional material has been included to provide a cultural perspective and to bring the reader up to date on recent treatment methods and legal applications which have evolved from the psycho-social approach to marital disorders.

The psychiatrist, in the course of treating a patient, can observe the unforeseen and disturbing changes of attitude toward the partner that come about in some marriages—the open and disguised manifestations of mutual provocation, the pyramiding of anxiety, the defensive withdrawals, and the deterioration in feelings and in behavior. Unless checked, these factors ultimately lead to actual separation or divorce, or to that stalemate best described as a chronic state of emotional divorce. As described in this volume, the psychiatrist tries to salvage such marriages, if at all possible, by exposing the neurotic components in the constant interplay of mutual hostility, i.e. in the compulsion to repeat forgotten infantile situations in the current relationship.

The widening clinical extension of psychoanalysis and the psychiatrist's greater role as consultant in casework agencies have also provided an abundance of data based not only on observations of the individual, but of both partners in a marriage, and often of their children as well. This development has led to the concept of family diagnosis and treatment advanced in these chapters.

Marital relationships are discussed in terms of diagnosis, prognosis and treatment, with most of the material presented in this sequence. The psychiatric chapters emphasize individual diagnosis; the casework material, derived from two large family agencies, stresses psycho-social diagnosis. The psychoanalytic contributions present documented studies on the fine interplay of unconscious emotions in marital interaction which can either disrupt a satisfactory marriage or neurotically perpetuate an unhappy one. Two chapters by research psychologists offer unique data obtained from projective tests in the study of both partners in a marriage.

While it makes considerable difference whether the partners we are dealing with are neurotic, psychotic, alcoholic, or relatively normal, cultural as well as diagnostic factors affect the types of expression of marital conflict. There are special sections devoted to these aspects.

The basic components of neurotic interaction in marriage have necessarily been discussed throughout the volume, albeit from different perspectives. The matter of neurotic choice of mates, for instance, is fundamental to the thesis of almost every chapter in this book, but it is given separate detailed consideration in one chapter. Similarly, reciprocal neurotic patterns, which are evident throughout the material, are the subject of a special section. Sexual relationships in marriage—affected in all modes of interaction described in this book—are systematically discussed in a chapter dealing with the most common forms of sexual disturbance.

The effects of chronic marital discord upon the development of children are evident in the case material showing how rejected or otherwise traumatized children reach adulthood and go into the world with feelings of inferiority and hostility and with a crippled capacity to love. The dynamic parent-child interaction is the focus of a chapter devoted to children's reactions to disturbed marital relations.

The practical application of the psychiatric and casework approach is covered by several contributions dealing with the evolution of newer methods of treating marital problems. Particular emphasis is given the family agency as a community service. While the individual psychiatrist can treat only a few patients a day, the family agencies he serves as consultant are now able to give much needed help to thousands.

Group therapy, one of the recent approaches, is described in a chapter which includes a discussion of group education and group counseling. The possibilities of predicting marital adjustment and the present limitations of the scientific method also receive careful consideration. Finally, a section on the legal applications of the psychiatric approach should prove useful in guiding those courts of law which deal with family matters.

While these contributions are representative of the most advanced thinking and practice in the field of marital counseling, they only lead the way toward a vast field that has yet to be intensively explored by social and psychiatric research. At present, the best professional opinion cannot prognosticate which marriages will be successful. Obviously, the psychiatric approach is only one of many avenues of investigation. It would be most desirable to integrate the findings in this field with those of anthropologists, sociologists, educators, economists and other professional workers in a wide study of factors affecting marital stability.

Toward this end, the World Federation for Mental Health, of which the Psychiatric Forum Group is a member, has invited its subsidiary associations to form interdisciplinary working groups to study *The Dynamics of Family Life.*

There is reason to believe that the scientific approach to problems of disturbed family relationships may shed some light on possible techniques for the study and solution of group, racial and national tensions. It is hoped that the contributions in this volume will serve as foundation stones for wider research projects which begin with a study of the family.

I wish to thank the sponsors of the Psychiatric Forum Group and its members for their cooperation in developing the studies on which this volume is based. I am deeply grateful to Drs. Paul Hoch, Sol W. Ginsburg, Leopold Bellak and Louis Linn for serving on the Editorial Advisory Board.

I would also like to extend a special word of appreciation to Dr. Marcel Heiman, past Chairman and long-time Secretary of the Psychiatric Forum Group, whose diligence and wisdom helped launch the stimulating symposia on marital problems at the New York Academy of Sciences.

I am very grateful to my wife, Helen, who provided many kinds of help in the preparation of this volume. I thank Henrietta Gilden for her valuable assistance with the editorial work. I wish to acknowledge with thanks the case material provided by Dr. Warner Muensterberger, and by several family service agencies.

I greatly appreciate the permission given by the editors of *The Psychoanalytic Quarterly, Social Casework* and *Texas Reports on Biology and Medicine* to reprint, with some editorial changes, sections of articles previously published in their journals. I would also like to acknowledge permission granted by the publishers to reprint in Chapter II of this work material derived from the following volumes: *The People in Your Life,* edited by Margaret M. Hughes, published by Alfred A. Knopf, Inc., and *Practical and Theoretical Aspects of Psychoanalysis,* by Lawrence S. Kubie, M.D., published by International Universities Press, Inc.

New York, 1956   *Victor W. Eisenstein, M.D.*

Contents

———————◆———————

College of St. Scholastica, Department of Nursing
Our Lady of Victory Hall, Duluth 5, Minnesota

NEUROTIC INTERACTION IN MARRIAGE

Marriage

A CULTURAL PERSPECTIVE

M. F. ASHLEY MONTAGU, Ph.D.*

THE READER who opens a book dealing with "interaction in marriage" expects logically to encounter a treatise on individuals directly involved in any marriage—the spouses and perhaps their children. The subsequent chapters of this book deal essentially with the behavior of these individuals, with the neurotic processes that influence and sometimes dominate the relationships between marriage partners. But the very perceptiveness and exhaustiveness that make these chapters valuable contributions to the study of human behavior need not divert us from recognizing that marriage involves not merely the individuals participating in it but also the society and the culture from which the individuals come and in which the marriage is situated. Neurotic interaction takes place essentially between individuals, but the nature of the individuals and of the marriage itself cannot be thoroughly under-

---

* Director of Research, New Jersey Committee for Physical Development and Mental Health; Lecturer in Anthropology, New School for Social Research, New York; Consultant to UNESCO.

stood without some understanding of their cultural and societal context.

In our twentieth-century, technologically advanced, Western culture, no less than in the more primitive cultures of peasant China or the Pacific Islands, marriage, both in its general characteristics and in its "texture," is a cultural product. In primitive and more stable societies, the cultural influences are readily apparent. In our own highly complex and rapidly changing society, they may be more difficult to perceive, but they are fully as significant. A brief review of the cultural influences on marriage in our own society may provide us with a perspective for the subject of this book—the interactions between the individual marital partners.

Perhaps one of the most striking characteristics of marriage in our culture is the fact that it is based upon the concept of romantic love. This concept of love, developed in the twelfth and thirteenth centuries among the nobility of France, was characterized by a complete abdication of all selfish motives, complete fealty and a complete idealization of the beloved. Love was held to be a matter of free exchange, and that which was freely given was conceived to be vastly superior to the dutiful relationship supposed to exist between husband and wife in marriages that were arranged by parents or overlords.

The concept of romantic love,* appealing as it did to women, has been handed down through the centuries and is now held by most women in our own culture—but not by most men. Romantic love does, of course, govern the male's courtship behavior. During this period he behaves much like the adoring lover described by the twelfth-century troubadours. But after marriage, the male—because he has never been culturally conditioned to it —cannot maintain the role of romantic lover, and the wife's disillusionment at the change in him is one of the causes of marital

---

* In a psychoanalytic frame of reference this unrealistic type of love is based on the re-awakening of the family romances of early childhood. See especially Chapters II, III and VI.

dissatisfaction. Such disillusionment does not occur, of course, in other cultures, in which romantic love plays no part.

Closely related to the romantic ideal is the notion that a principal function of marriage is to increase one's personal happiness. This view, too, is peculiar to Western culture and does not exist in those cultures in which marriage is arranged by parents or has essentially an economic basis. Since happiness is not likely to be achieved by purposeful pursuit, this hedonistic view leads almost inevitably to some degree of disappointment and disillusionment in marriage. On the other hand, since happiness is frequently a by-product of work well done and a sharing of interests, it is not necessarily absent from marriages in those cultures in which it is not regarded as a distinct element of marriage.

In almost every society, the division of labor between husband and wife is an important determinant of the stability and happiness of marriage. In nonliterate societies, this division of labor is clear-cut: the wife is usually the domestic worker, taking care of the feeding and clothing of the household and gathering the agricultural products, whereas the man is the hunter and the maker of implements. Each sex is trained from childhood in the fulfillment of its specific role. In our culture today, however, males and females are educated in virtually the same skills, and women are not educated specifically for the role of housewife. As a result, the woman acquires aspirations which are necessarily truncated after marriage; and her resulting frustrations are by no means lessened when she sees her husband able to pursue freely both the aspirations and the skills which he developed in the course of the same kind of education.*

The fact that the husband usually earns a living at a place remote from the household and often by means of a skill so highly complex as to be unsharable by the wife produces a bar-

---

* The "masculinization" of women and the corresponding "feminization" of men in our culture, and the activation of infantile envy and hostility are clinically related to many marital discord problems. See especially Chapters IV, IX, X, XIII and XV.

rier to communication between husband and wife and reduces sharply their area of common interests. If, on the other hand, the wife pursues her skills in an active career, the culture tends to make her feel guilty about forsaking her duties as housewife and mother.

The social mobility that is characteristic of our own culture is another source of tension in marriage. In our culture more than in any other, individuals marry outside their social class and find adjustment to the customs and values of the spouse's class difficult, repugnant, or even impossible. Moreover, even spouses of the same social class may find themselves rising to another level, and their joint difficulties of adjustment may drive them apart rather than bring them together. The majority of marriages are, of course, made between members of the same class, but so strong is the cultural pressure toward upward mobility that one spouse may perceive the other as his social inferior and may feel that his marriage prevented him from rising in the socio-economic scale.

Horizontal social mobility—that is, the movement of individuals geographically and occupationally without change in socio-economic level—tends to produce marriages with partners outside one's own regional, ethnic, or racial group—a practice virtually unheard of and socially proscribed in nonliterate societies. That differences between partners of such exogamous marriages contribute to marital instability is corroborated by considerable empirical data. A common example of neurotically motivated intermarriage is the marriage of an individual who feels rejected by his own group to a spouse belonging to a group that is considered in some way inferior. Often, too, exogamous marriages are simply neurotic manifestations of defiance toward parents or kinship group. But marriages outside one's own group are produced by the culture as well as by neurosis.

Perhaps one reason for the instability of exogamous marriages is that they lack the stabilizing influence of the tribe or kinship group. Marriage in nonliterate societies involves not only the two spouses but also the extended family groups of each of them. The multiplicity of interrelationships of each spouse with the other's kinship group promotes a stability that is often entirely lacking in the marriage of two essentially family-less individuals.

The relative ease of separation or divorce which is characteristic of our culture may be due in part to the lack of influence of the kinship group. But in its own right it can promote marriage instability by permitting a tentative approach to marriage. Where divorce is easy, two individuals can enter marriage with the realization that it is not an irrevocable step—indeed, that it may be entered into on a trial basis. Such an attitude toward marriage may, of course, be favorable for the mental health of the individuals involved, even though it produces statistics that bode ill for the stability of marriages. Our concern here is not with whether such an attitude is "good" or "bad"; rather, we should note that it is not prevalent in most cultures but is peculiar to our own.

The premium placed upon youthfulness and physical beauty is another peculiarity of our culture that can make for difficulties in marriage. The value of beauty is so strongly and so pervasively stressed that many men seem to make a marriage choice largely on the basis of being able to display their wives in accordance with Veblen's principle of conspicuous consumption. Marriage on the basis of physical attractiveness has had such disastrous consequences that society has been forced to permit the dissolution of such unions by making divorce more easily possible. The easing of divorce has, in turn, lessened the stability of marriage.

The sheer complexity of our society makes our marital roles much more complex than those in any other society. The wife must be competent not only as housewife and mother but also as companion and helpmeet—even though neither her personality nor her learned skills are suited to such a diversity of roles. The husband must be competent not only as a breadwinner but also as a *pater familias,* even though most of his waking hours are spent far away from the home on matters not even remotely connected with home or family, and even though he could not as a child learn his future roles by observing the activities of adult males. In nonliterate societies, by contrast, the growing child has immediate daily contact with every aspect of adult life. Family roles are unchanging and an adult can function successfully largely by following the parental pattern.

Virtually everything we have noted thus far would seem to indicate that marriage in our own culture is complex, difficult and precarious, and, by implication, marriage in nonliterate societies involves infinitely fewer tensions, difficulties and handicaps. In actual fact, very few studies have been made of marriage in nonliterate societies, but the systematic studies available and the incidental observations of anthropologists studying other institutions make it quite clear that in every society there are tensions, insecurities and difficulties. For example, sexual practices generally considered disgusting in one culture may find wide acceptance in another, and what is normal in one setting may actually be a neurotic deviation in another. For us, marriage is beset by a number of difficulties that are unknown in simpler societies, but it also offers infinitely more opportunity for the full development of personality and self-expression. In a stable society, marriage is likely to be stable, but it is also likely to be devoid of challenge and devoid of the satisfactions arising from the successful meeting of challenge. Moreover, the very complexity of our society permits deviations that are not possible in simpler societies, and as a result the deviant personality cannot only make a better adjustment but can contribute to the enrichment of the society itself.

In any human relationship stresses and strains are inevitable. The mentally healthy society is not marked by the complete absence of strain; rather it prepares its members to recognize the causes of interpersonal difficulties and to handle them in an effective and healthy fashion. Our own society, by its very nature, creates certain difficulties for marriage that are not to be found in simpler societies. On the other hand, we possess, to a degree not found in any other society, both the knowledge and the skills necessary for identifying and overcoming these difficulties. Our problem, then, is largely one of making available this knowledge and these skills and applying them more effectively. A first step is, of course, the identification and analysis of the tensions and difficulties of the contemporary marriage relationship. This is done with thoroughness and understanding in the chapters that follow.

Psychoanalysis and Marriage

PRACTICAL AND THEORETICAL ISSUES

———◆◆◆———

LAWRENCE S. KUBIE, M.D.*

The Role of Unconscious Factors in Marital Choice and Its Evolution

I SHALL ATTEMPT TO approach the problem of marriage on the basis of the following observed facts:

1. At the present point in the evolution of human culture a neurotic process is universal.

2. There are masked neurotic ingredients in every human personality.

3. Except in rare and fortuitous cases, these masked neurotic ingredients of the so-called "normal" play a distorting role in marriage choices.

4. Their role is inflated and reinforced by the romantic western tradition, which rationalizes and beatifies a neurotic state of obsessional infatuation.

5. The subsequent fate of the marriage depends too largely

---

* Clinical Professor of Psychiatry, School of Medicine, Yale University; Faculty, New York Psychoanalytic Institute; Lecturer in Psychiatry, College of Physicians and Surgeons, Columbia University, New York.

on the evolution of the masked neurotic processes which brought the couple together.

6. There is nothing new about this problem; it has been veiled for centuries because of the many marriages which were terminated by early deaths. It is now exposed by the steady lengthening of life expectancy.

7. This universal problem is now under psychoanalytic exploration, which is attempting to learn both how to avoid and to remedy it.

It is not possible to prove all of these points within the limits of a single chapter. Here we can only illustrate them. Nor is this discussion designed to offer any easy solutions. It will first attempt rather to describe a random sample of the problems that must be solved if man is ever to succeed in making of marriage the creative force that every sober student of the human comedy wants it to become. For there can be no question that divorce is always a tragedy, no matter how civilized the handling of it; always a confession of human failure, even when it is the sorry better of sorry alternatives.

The psychoanalyst studies divorce as the medical scientist goes to the autopsy table, there to learn from this consequence of masked human illness its causes and how to avoid their repetition. Just as medicine has learned from sickness and the autopsy table most of what we know of normal physiology, so it is from a study of marital failures that we must learn how to achieve marital success. Therefore, we will also indicate briefly the vital role that psychoanalysis must play in the study and solution of these failures.

Before going further, let me dispel one basic misconception: People think that the disruption rate of marriages is increasing. This is not true. In fact, quite on the contrary, it may be decreasing. According to the most accurate available statistics,[1] from 1890 to 1940 there was a drop in the rate of disruptions from about thirty-three disruptions per one thousand marriages to thirty. In that same period, however, the cause of disruption

changed. In 1890 about thirty disruptions per one thousand occurred because of death, and less than three per one thousand
were caused by divorce. In 1940, marital disruptions by death
had dropped from thirty to twenty-one per one thousand,
whereas disruptions by divorce had risen from three to nine per
one thousand. This is a drop of nine against an increase of six.
Thus the rising tide of divorces has not been large enough to
offset the decrease in family disruption due to the decline in the
death rate. The surprising net result shows that more families
were holding together in 1940 than in 1890.

The difference remains, however, that divorce today is accomplishing some of the reshuffling of marriages which only a
few years ago occurred through death. Fifty years ago death was
the rule during the age period between twenty and fifty. The
divorce rate has increased largely because of the increased life
expectancy in the middle years. Apparently, longevity has now
exposed for study the fact that the human race never has been
mature enough for enduring marriages, a fact which used to be
obscured by early deaths.

Just as the statistics for cancer and heart disease have risen
as we survive into the age period for these illnesses, so our divorce statistics are up because we live longer.* This increase in
divorce in the middle years accounts for some increase in divorce in the early years of marriage as an imitative phenomenon. It might be expected that young people who are brought
up in an atmosphere in which divorce is accepted as a deplora-
ble but not disreputable event, would be likely to turn to divorce more quickly than are young people who are brought up
in a more rigid atmosphere. Available statistics are inadequate
for the precise measurement of this tendency; but those that are
available seem to indicate that divorce by unhappy example is
not frequent enough to make a significant difference in the statistics.[2]

---

* See Chart 1.

CHART 1

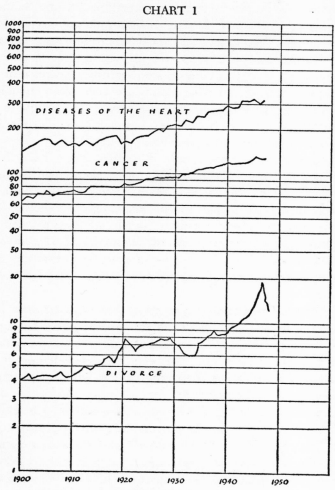

(Crude rates unadjusted for the aging of the population)
(Personal communication: Paul H. Jacobson, Vital Statistics Analyst,
Metropolitan Life Insurance Company; April 18, 1950)

I will digress for a moment to point out that, for adults at least, the psychological impact of a separation due to death may be quite different from the impact of a separation due to divorce. Small children may consciously resent death as though it were a willful desertion; but for normal adults death leaves behind it no sense of willful abandonment on a conscious level, and therefore no personal humiliation, no personal sting and no secret rage. There is only loneliness and longing.

For the normal adult, therefore, separation and loss through death are usually easier to bear than is separation through divorce. The impact on small children, on the other hand, is more complex; and much remains to be learned about the difference in the effects on children of losing a parent through death or through divorce. A detailed consideration of these questions is outside the scope of this discussion; but I could not refer to the shift from disruption by death to disruption through divorce without alluding to this example of the many unexplored psychological consequences of marital dissolution.

The fact that, in spite of all of the social pressures that oppose it, divorce increases as the death rate falls indicates that the sources of marital discord are ubiquitous. They are to be found in everyone. There are marriages that are intrinsically so sick that they should never have been contracted; because they were made not in heaven but in the neurotic component of the so-called normal human being.

I am not referring to frivolous and irresponsible marriages, nor to marriages that are the product of a quick infatuation. I am talking of marriages made soberly and in good faith between thoughtful and sincere people, but under the predominant influence of unconscious forces which impelled two individuals into an alliance that could not meet these unconscious demands. It will be my basic thesis that a major source of unhappiness between husband and wife is to be found in the discrepancies between their conscious and unconscious demands on each other and on the marriage, as these are expressed first in the choosing of a mate and then in the subsequent evolution of their relationship.

The only way to clarify this thesis is through example. Men and women are infinitely ingenious in their ability to find new ways of being unhappy together; so that even with unlimited space it would be impossible to illustrate every variety of marital misery. Therefore, I will restrict myself almost entirely to one aspect of the problem: namely, the unconscious forces that make it difficult for people to know what they are seeking in marriage, and how this confusion influences both the initial choice of a mate and the fate of the marriage.

CHART 2

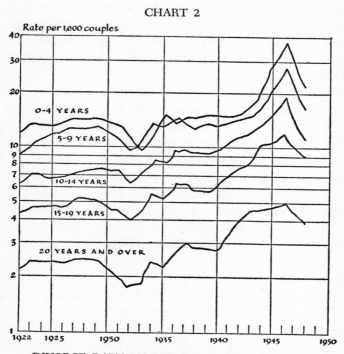

DIVORCE RATE BY DURATION OF MARRIAGE
UNITED STATES, 1922-1948
(From Paul H. Jacobson, in *American Sociological Review*, April 1950)

A young woman had been left fatherless at an early age. Over a number of years it was evident to all of her friends, but not to herself, that she was driven by an obvious need to find in marriage a substitute father and an ally against her mother. She ran through a series of engagements with older men, and finally married one who was within a year or so of her father's age only to discover that a substitute is never more than a substitute, and that in spite of the age discrepancy he wanted to be mothered as much as she wanted fathering. Since each felt cheated, they ended in a snarl of bitter recriminations.

Another young woman had always been dominated by a secret feeling that there was something wrong with her. Consequently, everything she ever did had concealment as its major purpose. She had been engaged many times to wholly suitable young men; but had to break each engagement to avoid the exposures of physical intimacy. Finally, she was able to go through with a marriage to the suitor she cared *least* about, and then only because she cared so little for his opinion, and because it meant living in a distant and obscure land where she felt hidden. She did not realize this until a last-minute change in his plans threw her into a turmoil of panicky indecision.

Needs that are obscure to the man or woman often take forms that are transparent to everyone else. One wholly unintellectual youngster married a college professor in an effort to triumph both over her own personal sense of degradation and over her mother. Another, the intelligent and highly educated daughter of an intellectual professional woman, married a famous athlete of little brain in an attempt to enable matter to triumph over mind and mother.

Another girl of beauty, intelligence and charm was compelled to choose an ill-favored, homely, poorly equipped, neurotic and dependent man in an unconscious effort to hide herself. The discrepancy between the two shocked all her friends. But just as the Rorschach and the Szondi tests represent in part each man's unconscious image of himself, this unfortunate man represented this girl's unconscious distorted image of herself. In her choice of a husband she was saying to the world: "This is all that I am, and all that I am good for."

A strange yet frequent manifestation of the role of unconscious forces in the choice of a mate is seen in those instances in which a youngster chooses someone who looks like himself, but also in those related instances in which the choice is of a diametrically opposite type. These are not accidents. They depend on the balance of unconscious self-love and self-hate in the structure of each personality. Recently, I encountered a young couple who had worked out an excellent compromise, combining resemblance and contrast. They looked enough alike to have been twins, but one was very dark of hair and skin, the other strikingly fair.

A patient's mother was a vigorous, dominant, aggressive woman. Her father was technically able, but emotionally weak, insecure and colorless. The patient was aware of this discrepancy and hated both parents for it. Before a class we predicted correctly the kind of marriage to expect. The young woman felt compelled to choose someone in her father's image, so that in her own marriage she could be as dominant a figure as her mother had been. Quite unconsciously she had to duplicate in her own life everything she had hated most in her mother's behavior toward her father. The ultimate outcome was equally predictable. She would hate the man when he submitted just as she had resented her father's weakness. Yet, with the image of her mother in her heart, she would be forced automatically to fight him violently if he opposed her. Inevitably, this split ended in illness.

Two youngsters had grown up insecure and lonely, seclusive and bookish in their tastes, in various ways apart from the general run of adolescents. They drew together through mutual sympathy and the compatibility of their intellectual and artistic interests, through their understanding of each other's needs and problems, and in some measure just because misery loves company. During their courtship the sense of loneliness was gone. Each had an ally. Almost for the first time each had some place to go, and someone with whom to share life. Each was literally all the world to the other.

Unfortunately, when they married they discovered that when they faced the world together something quite unexpected happened. They could no longer be the whole world to each other. Instead, they had to reach out to the world to bring it into their joined lives. Yet each was still frightened of this outside world. In a sense each pushed the other, saying "You go first." But neither could. So they became angry at each other. The separate and private misery that each had brought into the marriage and that originally had drawn them together was now compounded in a marriage that had been contracted in an unrealistic expectation that it would lessen that misery. Instead, in marriage the unresolved neurosis of one was added to the neurosis of the other. That social shyness which had united them became something hampering, which one resented in the other, and which ultimately drove them apart.

They were learning bitterly and painfully a lesson that humanity as a whole has never learned—that no one has ever married himself out of a neurosis. Instead, when two young people are drawn into marriage by the lure of the other's illness, one will add the weight of his own neurotic infirmity to that of the other, with growing pain and resentment.

This confusion in the choice of a mate, and the unattainability of the unconscious purposes that determine this choice lead to unexpected and sometimes unpredictable changes after marriage, which may develop gradually or suddenly. An Englishman with active gregarious impulses had had a long and lonely battle with tuberculosis. During his recovery he met a girl with whom he shared many interests, among them a warm interest in human beings. Indeed, he was first drawn to her by the range and warmth of her friendships. After marriage an unforeseen change occurred. Neither of them had realized that her sociability was possible for her only as long as she was a bachelor girl, and that the moment she stepped into marriage her gregarious impulses would become so painfully inhibited that even with old friends she would become tense and awkward and silent. They had no way of foreseeing that she could be happy among other human beings only as a soloist, never as part of a duet. This was a streak of illness in her that could come to light only after marriage.

Indeed, without her knowledge, this had been why she had several times shied away from marriage to compatible men at the last moment. On this occasion, it had been only because her husband was still ill, and therefore socially inactive, that she had been able to marry him. They did not realize any of this until after marriage, when almost overnight she became as misanthropic as she had previously been warmhearted and outgoing, forcing her husband back into the loneliness and isolation in which his tuberculosis had imprisoned him for years. This destroyed the rapport between them. They became deeply unhappy; yet for a long time could not understand what had happened. He could never induce her to seek treatment, with the result that ultimately they separated. It added greatly to his subsequent bitterness to discover that immediately after this separation she became again the happily gregarious person who had originally attracted him.

In another situation, the shoe was on the other foot. Both were gregarious persons with innumerable friends. The man married a girl of beauty, wit and intelligence, who was in his own profession. He looked forward eagerly to taking her with him to meetings and sharing with her the many activities in which he engaged. She came from a small and distant town, where she had made a name for herself, and she too looked forward to sharing their interests.

It did not work out that way. Instead, he found himself making excuses to avoid taking her to meetings, or to avoid going at all. If he went with her he carefully avoided her, or became inexplicably sulky, silent, tense and awkward. Fortunately, they had the wisdom to seek treatment before the relationship had become too strained for help. He found out why after marriage he had become unable to share any aspect of life with a woman with whom he had shared everything during courtship. This automatic reversal was a manifestation of an unsuspected neurotic conflict, which through its various manifestations would have destroyed the marriage if it had been allowed to go untreated.

Another unanticipated post-nuptial change often occurs when an aggressive woman marries a man to dominate him without realizing what she is doing. I have in mind one such woman who in manner, clothes, build and voice was gentle and tender. Yet, because of childhood hurts at the hands of her older brothers, she had an inflexible hidden determination always to hold the upper hand in any relationship with any man. Unconsciously, she had always chosen friends on that basis.

When she finally married, she and her husband found themselves in a dilemma. He was not weak, but he was a sweet and kindly man, always eager to do things to please her. After marriage they gradually became aware of a change. Without her realizing it, her victory had become empty, and she became restless. Unwittingly, she was seeking new men to conquer. When he realized that she not only took his submissiveness for granted but scorned it, he reversed his role and stood up to her. This was equally intolerable to her. Soon the marriage was at the breaking point and they had to seek help.

A variation on this theme was experienced by the oldest of five sisters, a girl who throughout her childhood had held a dominant position in the family because of her athletic skill, her beauty and her great intelligence. Her successes gave her no sense of satisfaction, however, because they never enabled her to realize her unconscious goal, i.e. to be her father's oldest son and thus to become the head of the family. She had to play out this goal in life by always choosing a man who in one way or another could be stamped as inferior. The same need shaped her courtship; but when the relationship was established as an engagement and finally as a marriage, the pain it caused her to be publicly identified with an inferior man turned her love to hate, her admiration to scorn, and ultimately destroyed the relationship.

Another woman, who had been left motherless at an early age, mothered her two younger brothers and her older brother as well. She hid all rivalry in this maternal quality. This made her lovable, but incapable of treating any man as an equal. Throughout her life men could never be anything to her except children, i.e. her young brothers. This determined her choice of a husband, and also in part the subsequent deterioration of her marriage.

Interestingly enough, the same set of circumstances had a comparable effect on her older brother. It determined his choice of medicine and of gynecology as a career, and made it impossible for him ever to love a woman who was not sick or in trouble.

Many marriages illustrate the fatal discrepancy between conscious, attainable needs and unconscious, unattainable goals as a major source of marital discord. A widely prevalent example of this is seen among the many men and women who marry with the major unconscious purpose of finding a parent. This may take varied forms. Unconsciously, the woman may be marrying not her fiancé but his father or mother; the man may be marrying his fiancée's mother or older sister or aunt, or, for that matter, her father.

During the courtship, while the marriage is still in the offing, any vague feelings of discontent or incompleteness will be balanced by the reassuring hope that fulfillment merely waits on the next step—the marriage itself, or a new home, or the advent of a child or of ten children, or more money, and so on. After marriage, however, the time comes when all such milestones will have been reached without dispelling the feelings of

emptiness and unfulfillment. If we keep in mind the unconscious and unattainable goal of the marriage, then this discontent is almost mathematically predictable. It has nothing to do
with the potential compatibility of the couple's interests and
standards.

As an example, I think of a woman who had been deeply
hurt by her own father. He had rejected her completely, turning all his affection on her brother. She married a man of whom
she was deeply fond. On the surface this was because of his fine
qualities, but unconsciously she chose him because she needed
his father to replace her own. Before her marriage her prospective father-in-law had paid her a great deal of loving attention.
She had not been married long, however, before she discovered that this had been only because of his passionate devotion
to his son, her future husband, whom he had courted indirectly by courting her affection. Thus, the ultimate situation of
her marriage duplicated with fantastic and tragic fidelity the
situation of her childhood, with father-in-law and husband in
the roles previously played by her own father and brother. In
spite of her devotion to her husband, this rejection by her father-in-law whipped up in her the old hatred of her brother,
turned now in blind fury against her husband. Had she not
come for help, this would surely have destroyed her marriage.

Sometimes, because of a similarly unconscious need for a
mother, the man may try to make a mother of his wife; while
at the same time the wife will try to turn her husband into a
father. Thus two immature people of middle age chose each
other because of their immaturity, two babes in the woods at
forty. They married only to find that each needed and wanted
not a spouse but a parent. Although there is always a certain
amount of maternal and paternal feeling in any relationship between a man and a woman, when this simultaneous and conflicting demand becomes the predominant unconscious goal of
the marriage, each rejects the parental role into which the other
is trying to force him; and each thereupon feels hurt and re-

sentful without knowing why. Such a marriage inevitably ends in deep trouble.

The unconscious need to reproduce something out of the past or to wipe out an old pain can influence marriage choices in many destructive ways. I think of two instances of young men who married their sisters' best friends, believing and feeling that they were in love with their young brides. They were indeed in love; not, however, with the girls they married, but each with his own sister.

I think of another young man who was deeply attached to his sister but also intensely hostile to her, and for many reasons. Without knowing what he was doing, he fell in love with a girl and married her, not because of any essential compatibility between them, but solely because she was in every respect the opposite pole to his own sister. It was his way of getting back at the sister who had hurt him—hardly a sound basis for a happy marriage.

This need to wipe out an old pain or pay off an old score comes up again and again. A young woman who had been the only daughter and the youngest child in a large family grew up to be an over-conscientious, thoughtful, responsible and considerate youngster, but socially tense, awkward and insecure. Her brothers, on the other hand, were charming, alcoholic wastrels who had made her childhood and adolescence miserable. What did she do? She fell in love with and married a man who was one of her brothers' best friends and who shared all their faults. She married him in an unconscious effort to wipe away the pain that these brothers had caused her through so many years. He, on his part, married her out of an unconscious homosexual attachment to her brothers. The pitiful outcome was of course predictable; over the years he continually put her through the same kind of hurt that had scarred her childhood.

Not long ago, I saw a young man who had married himself into a predicament that was most painful. He had been an illegitimate child. Shortly after his birth, his mother had married a man with whom she subsequently had a large family. My patient had been brought up as part of this family and was deeply devoted to his mother and his stepfather. Like an automaton marching unconsciously to his doom, he tried to create in his own marriage a replica of his childhood by marrying his mother's image. That is, he married a girl who, like his mother, had had an illegitimate child by another man, seeking to find again in his marriage the warmth and affection of his infancy and childhood. Unfortunately, it did not work out that way for him—the girl proved to be a tramp. This was as though his own mother had betrayed him, and the hurt threw him into a deep depression.

There are men and women who show the domination of other types of unconscious purpose in their choices of mates. I have seen men who repeatedly choose alcoholic or frigid or unfaithful wives; women who repeatedly choose brutal, alcoholic, impotent or unfaithful husbands. In many of these women there is the history of an alcoholic father, and on analysis it becomes clear that the unconscious goal of the marriage was to cure the father's alcoholism in effigy, thus winning back the lost affection of the alcoholic father, while proving at the same time that she could do a better job than her mother.

Literature is full of tales of men who repeatedly choose women who will hurt them, i.e. women who cannot love a man. This is always in the service of unconscious needs. For instance, in their marriages human beings may try without knowing it to prove something about themselves that they themselves doubt obsessively. Thus, men may try to prove that they can win love from an unloving mother or older sister, represented in the unloving wife they have chosen. Or they may have to prove their potency or that they are not physically repulsive by overcoming the grim aloofness of a woman who is essentially hostile to all men.

We do this consciously in other aspects of life. It is inevitable that it should happen in marriage too. After all, if we have to prove that we are great mountain climbers, we climb Mount Everest, not the Berkshires. And whenever men and women have to prove that they are lovable they can feel that they are achieving their goal only if the attempt is made in the face of great obstacles.

The unhappiest feature of this need to prove something is its insatiability. Whenever the source of any self-doubt is unconscious, the need to prove it becomes endlessly repetitive. The simplest illustration of this is a hand-washing compulsion. The individual who suffers from such a compulsion knows that he does not have dirty hands. Nonetheless, no sooner does he finish scrubbing his hands than his secret doubts start welling up again. Within a matter of hours or even minutes, the compulsion to wash his hands again becomes overwhelming.

The same thing occurs in the relationship between a man and a woman whenever a need to allay doubts whose roots are unconscious is the dominant force in the marriage. Under these circumstances the relationship gradually becomes incapable of satisfying the necessity to prove, irrespective of the essential quality of that relationship. Just as Joe Louis could not go on believing that he was the heavyweight champion of the world if he continued giving exhibition matches with the same stumblebums, so a man with unconscious doubts about his potency cannot allay them by making love to the same woman all the time. Success with one woman is not enough. It no longer seems to count. He wonders whether he could succeed with anyone else, and ultimately his neurotic need to prove himself drives him into promiscuity. The same struggle with self doubts has driven many a woman into a like path. Many a man with the reputation of being a Don Juan, many a woman who is known as a girl-about-town, is merely a poor unfortunate soul running like mad with the devil of self-doubt at his or her heels, finding momentary surcease through promiscuity.

These self-doubts can be as unreal as a phobia of butterflies. The woman may be beautiful or the man a sexual athlete, and still they may be afflicted by these fears. On one occasion a young football star, a towering fellow with a magnificent build who had never in his life been impotent, came to me in tears because his home was about to break up. He was in love with his wife, but he was haunted by a fear of impotence which drove him against his will from woman to woman. This was merely an exaggerated example of something that in subtler form plays a role in many marriages.

It is hardly necessary to say that complicated forces that center on the problems of sexual adjustment can be important both in determining the choice of a mate and also in the subsequent evolution of a marriage. It is necessary, however, to make it clear from the start that an orgasm is not a panacea for all marital woe, and that sex can cause as much trouble when intercourse itself is physiologically successful as when it is unsuccessful. I have never seen a marriage made or broken by sex alone, except in the case of frank perversions.

I repeat that it is a gross oversimplification of this complex problem to think that the achievement of an orgasm marks the end of all difficulties. Paradoxically enough, it can sometimes be the start of trouble. There are individuals who bring into marriage such deep-seated feelings of sexual guilt that they can tolerate sex only as long as it is unsuccessful and consequently react to orgasm with guilt or panic. I have more than once treated a man who was on the verge of leaving his wife, with whom he was physically happy, because each orgasm threw him into a panic and from panic into rage and from rage into depression. I have dealt with women for whom an episode of happy and successful love-making always terminated in depression, followed by a subtle undercurrent of resentment that would express itself on entirely unrelated matters in the subsequent days.

Sometimes this leads the man to choose a frigid woman, or the woman to choose an impotent man. Each of these unhappy choices brings in its train its complicated frustrations. The woman's frigidity (her inability to have an orgasm) often means to the man that there is something defective about him, something inadequate in him as a lover. This too can arouse anxiety, resentment, depression and sometimes hypochondriacal fears of physiological inadequacy. Sometimes, it drives the man to attempts to prove himself with other women, not realizing that he may have chosen his wife precisely because she was frigid, and that he will run away from any woman who is sexually responsive in spite of his need for the reassurance such a woman could give him.

The woman who unwittingly always chooses impotent men also finds herself between Scylla and Charybdis. I think, for instance, of an extraordinarily beautiful woman whose secret image of herself was as homely, repulsive and, in some mysterious way, inadequate. Her ideal male was a huge athletic hero about whom she dreamed frequently. In real life, however, she turned only to diminutive men beside whom she felt less inferior. She ran from one to another in a vain attempt to find reassurance.

Another woman in a similar situation married an impotent man and then was tormented by the feeling that his impotence was her fault, thus confirming her horror of her own body. For a time this drove her into a life of promiscuity which she abhorred, and which depressed her deeply. Each erotic episode gave her the same momentary relief from unbearable pain that a man with a hand-washing compulsion gets from washing his hands. I would emphasize, however, that in both cases the relief is only momentary. Within a few minutes the torturing self-doubts start up again.

Paralleling the woman who lived out her fantasy-life with huge athletes and her real life with little scholars, there is the man who because of his own unconscious feelings of genital inferiority had to choose a small woman, or a woman with a small mouth and small breasts, with a dim notion that there would not be such a discrepancy in size between them. Then, however, he felt publicly humiliated, as though he had proclaimed to the world his genital inadequacy.

There is also the familiar yet perplexing phenomenon of the man who loses his potency with any woman he loves and respects, but who can be potent with one he scorns. I had a patient who had had a happy affair with a woman for many years, during which his wife was hopelessly ill. He married this mistress eagerly after his wife's death, only to become impotent with her on their wedding night and thereafter. Comparable reactions occur in women as well; in anyone, in fact, in whom the taboo on sex is so deep-seated that it can be shared only extra-legally or in the gutter.

Subtly or grossly these distortions occur with great frequency. They are not rare oddities. They occur because we grow up with a profoundly distorted attitude toward the human body. It is indeed remarkable that although we have been human beings for quite a number of years, we still do not accept the anatomical differences between the body of a man and the body of a woman. It is not accepted in art. Anyone who goes to the galleries can observe that a part of the artistic drive is a subtle effort to disguise and distort and alter the reality of the body—either by minimizing these differences or else by burlesquing them. Through the ages, we find the same thing in fashions both for men and women.

Anatole France in *Penguin Island* pointed out that when the penguins turned into human beings they lost their virtue only when they put on clothes, that is, when they hid the rejected realities of the body. Man is like this; and no young man or woman comes to marriage without bringing some measure of this deep-seated problem. Marriage demands that he overcome his secret rejection of his own body and that of his partner. This he cannot always do. Not infrequently, a compromise forms in his mind, such as an unconscious fantasy of finding some mysterious middle sex, one that is neither male nor female, or else both in one. We see this in the effeminization of men's fashions and the masculinization of women's.

The inability to accept our bodies as they are influences the choice of the marriage partner in a quite characteristic way, frequently forcing the young man or woman to choose someone who resembles himself or herself, as though each were marrying his own alter ego in the opposite sex. Here again, the unconscious goal can never be attained in actuality, and the symbolic act of marrying one's own counterpart in the opposite sex leads inevitably to bitter and seemingly inexplicable disillusion. The same confused rejection of the body also expresses itself in those labored variations on techniques of intercourse which fill the pages of books on the sexual act.

Most perplexing of all is the fact that a marriage that starts under completely happy auspices can sometimes go to pieces because of that very happiness. There are situations in which a man or woman has been so hostile to a parent that he or she cannot tolerate resembling the parent. This can occur even when the parent's marriage has been happy, if the child has felt excluded from it. As a consequence, even his own happy marriage can be intolerable to that child when he is grown up, because it forces on him an identification with a hated parent. This starts a deep surge of tension, rebellion and self-hatred. The hate that had originally been directed toward the parent now becomes self-hate; and the result is a paradoxical depression as a reaction to a happy marriage, with ultimate rejection of the marriage itself.

There are also the many choices which are made out of insecurity: the man who is too anxiety-ridden to stand on his own feet and who marries for external security of one kind or another; the woman who must have a prominent and successful husband, but not a busy one (that is, the woman who wants to be married to the President of the United States, but only while he is on vacation).

Finally, there are the competitive unions: the man who

chooses beauty, but cannot tolerate his wife's attractiveness and popularity; the woman who chooses success, but cannot tolerate her husband's career or even his physical prowess. For instance, one woman who was an intensely jealous rival, but who could never let herself express any of this rivalry in open competition, married her gladiator. That is, she married a highly successful, highly competitive businessman-athlete, and then could never share or enjoy any of his triumphs, but belittled his victories and compulsively made fun of him whenever his dexterity failed him.

Thus the difficulties of maintaining a happy relationship may start the moment the period of courtship is over—indeed, the moment uncertain pursuit has turned into victory. Two people who have been completely happy during a thoughtful and serious courtship can become equally unhappy once this relationship is established in marriage.

I hope I have made it clear that the choosing of a mate is one of the most confused steps a human being takes in life, and this not primarily because he chooses a mate whose interests and habits are incompatible with his own, but because each of the pair is ignorant of the unconscious purposes that determine their respective choices. This is why hasty and impulsive unions may stand up as well as those which have been made with the greatest possible conscious foresight. Both miscarry whenever unconscious goals exercise a preponderant influence which has been left out of account and which is at variance with conscious and preconscious goals.

Certainly, as our examples have shown, there are many marriages, made slowly and on the basis of compatible conscious interests and harmonious conscious goals, which should have been sound but which went on the rocks as the unconscious goals diverged. Those very activities which before marriage had been shared happily can unexpectedly become after marriage the source of bitter unhappiness. Indeed, so subtle are these changes that the quality of the courtship may be no index of what the marriage will be like.

Furthermore, it is obvious that the state of being in love is no guide. It is an obsessional state which, like all obsessions, is in part driven by unconscious anger. The transition from that strangely ambivalent obsession which we call "being in love" to the capacity really to love another human being is one of the most important and difficult and intricate phenomena of human life. Yet it is only in recent years that it has been subjected to critical scientific study.

All of this brings me to the conclusion that one of the fundamental challenges confronting us today is to discover how human beings, young and old, can be taught to distinguish between their conscious and attainable goals and needs on the one hand, and their unconscious and unattainable goals on the other. Until we can do this, the problem of human happiness, whether in marriage or otherwise, will remain unsolved.

This is a large order. Perhaps the answer will come through a basic and fundamental change in our entire system of education, a change which recognizes that no matter how well a human being is educated in chemistry, physics, economics, history or literature, he remains a barbarian unless he knows something about himself. Self-knowledge in depth, which has been the forgotten factor in our educational system, must become instead its primary focus. Ever since Socrates, the ideal of knowing ourselves has been recognized. Psychiatry has only added a deeper understanding of what knowing oneself implies. Like good intentions, self-knowledge is of little value unless it penetrates to the unconscious levels of the human spirit. On this depends the future of marriage.

Psychoanalysis, Marriage and Divorce

Psychoanalysis can sometimes make a significant contribution to the solution of marital problems, by uncovering sources of unconscious hostility, guilt and fear, and by clarifying their disguised influences in the marriage. In this way analysis can often lessen or even eliminate their destructive effects. Psychoanalysis may well be able someday to contribute a great deal to the prevention of marital problems; but only after young folk will have learned that they can never marry themselves out of their neuroses, and that whenever two unhappy young people marry, each adds the problems of his own neurosis to those of the other. From these experiences comes the hard but inescapable conclusion that we must earn our right to marry by solving our individual problems first.

Successful psychoanalytic help for a sick marriage frequently requires the conjoint action of the husband and the wife. All too often, however, before they reach the analyst's office the hostility has become too great for any such mutual effort. The situation is ideal where both come for help, but this is still the exception. More often one or the other comes secretly, or in fear, or in defiance of the other's wishes. Frequently, patients come to an analyst in a state of concealed panic, without having the slightest idea that their marriages are threatened. Unaware of what they are doing, they may try to strike a bargain with the psychoanalyst: "You may analyze me, if you will promise not to uncover any trouble in my marriage." Or they may come with an opposite purpose, trying to blame their own neurotic unhappiness on the marriage, and seeking a pseudo-psychiatric, pseudo-psychoanalytic justification for promiscuity or divorce.

At times, someone comes to seek help in a crisis caused

largely by the illness of a husband or wife who refuses to acknowledge either the illness itself or any need for treatment. The partner's symptoms may be as obvious as chronic alcoholism, impotence, frank sexual perversions, or sadistic compulsive states, yet he may still refuse to acknowledge his need for help. The patient may have shielded the other for years, so successfully that not even their most intimate friends or relatives have been aware of any difficulty, until the increasing strain and the incessant threat to the welfare of the children finally compels action. In such situations it is not unusual for the healthier of the pair to ask to be analyzed, hoping that by some trick of super-adaptation, some magic of absent treatment, the sick one may be cured without ever submitting to therapy.

Some patients come for analysis resentfully and only under the threat of impending divorce or separation; while others may seek help earnestly, but in the face of bitter resentment and neurotic jealousy on the part of the wife or husband.

These are some of the many and varied snarls which the analyst must untangle before he can decide which one of the partners most needs analysis, and which one, if either, can be analyzed. Sometimes when confronted with a marriage crisis it is hard for the analyst to decide whether or not to undertake an analysis at all, since no matter how successful the analysis may be for the one who accepts his own need for treatment, the outcome of the marriage will depend at least in part upon the ultimate attitude of the unanalyzed partner as well as upon that of the patient. Healthy adaptations can never come exclusively from one side.

It is evident, therefore, that marital problems frequently confront the analyst with difficult decisions. He knows that until a neurosis has hurt a patient there is little hope of his seeking therapy. We do not ask to be treated for those of our neurotic traits which hurt someone else while leaving us relatively unscathed. Indeed, where a neurosis expresses itself by causing pain in a marriage, the greater the pain it causes the more obstinate may be the self-justifying defenses which the potential patient will erect. Frequently, therefore, when an analyst faces a severely threatened and distorted marriage he may have to clear the air both for himself and for his patient by insisting upon a temporary separation before undertaking the analysis. Such a separation, when undertaken for purely technical reasons, can greatly facilitate the progress of an analysis and may actually save a marriage. On the other hand, this cannot always be the outcome. Sometimes the separation makes it possible for each partner to turn his back on his own contributions to the unhappiness of the marriage. Thus the analyst is in something of a dilemma. On the one hand, if the couple remains together, the constant interplay of mutual hostility is likely to swamp the analysis. On the other hand, separation may bring to each a false peace, in which the impulse to search into himself may be completely eliminated. It is impossible for the analyst to be all-wise in resolving this problem as he approaches the treatment of either member of such an unhappy marriage. The only thing he can do is try to maintain a status quo until he becomes aware of the more important of the unconscious forces which play below the surface.

Occasionally, however, even in situations which at the outset look hopeless, the gradual improvement of one of the partners may lessen the opposition of the other until he finally seeks help for himself as well. If the children have been psychologically injured in the family battle, they too may ultimately need treatment. This spread of psychoanalysis through a family often arouses cynical comments from onlookers. Nevertheless, to many a family this has meant the difference between happiness and divorce, between disturbed children and their harmonious growth and development.

It must be remembered, however, that psychoanalysis is a medical therapy, and as such must aim primarily at the preservation or restoration of mental and emotional health. The psychoanalyst is not a marriage broker, nor a marriage saver, nor yet a marriage wrecker. Whenever he faces a choice between health or marriage, his medical duty is at least to present this issue clearly, so that the patient can make his choice in the full awareness of its implications. Happily, such an impasse arises less frequently than is popularly believed.

People in general have an exaggerated impression of the frequency with which psychoanalysis ends in divorce. Marital conflicts, like other neurotic symptoms, are usually hidden as long as possible. As a result, although the caldron may have been boiling long before it was brought to the psychoanalyst to be cooled off, no one except the husband and wife may have any suspicion of the extent of the maladjustment. And when the couple finally turns to analysis, the breach may have become too wide for successful analytic intervention. This is one of the reasons for the impression that analysis regularly ends in the disruption of a marriage. Another reason for this impression is that when the analysis succeeds in resolving marital difficulties, no one may have known either that the marriage had been threatened or even that the husband or wife had undergone treatment. Therefore, one hears gossip about the analyses which end in divorce, but not about the successful analytic efforts to avert divorce. Divorce has the greater news value.

On the other hand it is true, of course, that divorce sometimes is inevitable. This may be due to any of several reasons:

1. Before the problems were brought to a psychoanalyst, the relationship between husband and wife may have been strained for so many years that even the restoration of health may not be enough to free the patients of their bitterness. Divorce may then be chosen in the end, despite the successful analysis of the neuroses which had destroyed the marriage.

2. Furthermore, many marriages are contracted not on a basis of health, but on that of neurotic purposes. The story is all too familiar of maladjusted young people who try to escape their separate miseries by joining their problems in marriage. Since marriage never cures a neurosis, it usually ends up by be-

ing blamed for it; and presently the neurotic angers of both partners are pitted against each other in a merciless battle. Even if this battle can be checked by analysis, the couple may find that with the elimination of the original neurotic reasons for the marriage, no healthy reason remains for continuing together. In such situations separation may be the only healthy solution.

Therefore, divorce cannot be used as an index of either the success or failure of an analysis. To do so would be tantamount to claiming that the function of psychoanalysis is not primarily to cure people of their neuroses but to enable them to put up with all the neurotic mistakes they have ever made. The psychoanalyst aims to save a marriage wherever this is humanly possible; the frequency of his success bears testimony both to his purpose and to his ability. Where, however, he must choose between a patient's health and a patient's marriage, this issue must be squarely faced. One cannot leave this topic without looking into the future and asking what would be the effect upon marital happiness and stability if young people had an opportunity to resolve their neuroses before they contracted neurotic alliances. This is a question for the coming decades to answer.

Of the technical problems which arise in the psychoanalytic treatment of marital difficulties, only two require elucidation here. The first is the problem of the concurrent analysis of husband and wife. This may be of great value when the two analyses are conducted independently by two psychoanalysts; but experienced psychoanalysts generally regard as unwise the simultaneous treatment of both husband and wife by one analyst. This makes no difficulty for the analyst; but, contrary to usual expectations, it makes the task of both patients harder. In the end one or the other is likely to lose his confidence in the impartiality of the analyst; and the analysis of this patient will suffer accordingly. Therefore, since it is usually wise not to postpone the analysis of one until after the other is finished, it is well to send the second patient to another psychoanalyst as soon as possible.

On the other hand, in the preparatory antechamber to the analysis, or in cases of marital maladjustment which are not to be psychoanalyzed, both patients can sometimes be handled effectively by the same physician. Yet even here, if the currents of hostile feeling run strong, it may be advantageous to seek the help of two physicians instead of one.

In the analysis of marital problems, one other technical problem is of practical importance, namely the problem that arises whenever a patient has to work out both a severe personal neurotic problem and a complicated marital conflict. In such circumstances it may be extremely difficult to make headway without separating the two problems. In order to do this, it sometimes is necessary to arrange for a temporary separation, so that the individual's neurotic difficulties can be clarified in some measure before tackling the problems of mutual adjustment. Furthermore, in any analysis there may be periods during which neurotic rage is stirred up; and it is almost inevitable that this rage will be vented on the marital partner. A technical separation during part of an analysis may spare a marriage stresses and strains which might otherwise do final and irreparable damage to it.

The Relation of Psychoanalysis to the Prevention and Cure of Marital Problems

The illustrative material which has already been presented will indicate that the problem of marriage is identical with the problem of the universal neurosis. Therefore, the discussion of the role of analysis in relation to marriage cannot be approached as something distinct from the problems which are involved in psychoanalytic psychotherapy in general. Especially

relevant are its early use in childhood in an effort to prevent
the cumulative effects of early neurotic episodes; its use in late
puberty to produce better integrated and more mature adolescents; and its use in adolescence to resolve the neurotic problems which the teen-ager so often attempts to resolve in his
adolescent infatuations, but which he carries over into his
young adult years where they determine marriage choices and
the fate of marriages. In other words, all of the problems which
are involved in the early use of psychoanalytic knowledge and
of psychoanalytic therapy for preventive purposes are part of
the problem of how to use psychoanalysis as a technique to
facilitate mature marriages.

Psychoanalysis in marriages which are already on the rocks
almost always encounters one particular complication. In marriage individuals usually come not to get well, but to prove to a
spouse that they *are* well; not to find out where they are wrong,
but to prove they are right. Usually, they come refusing to accept the responsibility for their own neurotic pain, and eager to
blame it on the other. Therefore, the use of analysis for the
resolution of marriage problems starts under a handicap, with
which it does not have to contend when individuals come because of the emotional suffering caused by their neuroses. All
too often in marriage the neurosis of one partner causes pain
chiefly to the other. The drive toward health, the willingness to
sacrifice on the path to health, and the willingness to accept
the pain of treatment are much weaker under these circumstances than they are in the patient who is suffering because of
the nature of his own neurotic symptoms.

In other words, in the treatment of marital problems psychoanalysis tends in general to have a lesser leverage than it has
when dealing directly with an individual's neurosis. In view of
this, it is remarkable that analysis succeeds as often as it does in
salvaging, preserving and in maturing the relationship between
the married pair and between the parents and the children. To
date, there are no statistics on marriages saved or marriages lost

through psychoanalysis; the American Psychoanalytic Association is currently engaged in a research project on this point. Informal inquiry indicates, however, that the score is surprisingly good.

These many considerations can best be focused by considering a few key questions:

Since human beings are imperfect, since few of us have insight into our unconscious purposes when we make our marriage choices, since few of us are emotionally mature when we marry, and since there often are unanticipated changes between the choosing period and the living period, how can we be sure of making a right choice? How can we guide young people or advise them? Are there any ways of testing, any sure indices?

In the present stage of our knowledge we have no sure guides. Our position here is comparable to that of medicine in relation to many illnesses. We can diagnose the existence of an illness, and even know something of its nature, long before we are able either to prevent or cure it. The same thing is true of the problems of marriage. We have no indication that a battery of psychological tests would help us. Furthermore, apart from the fact that it would be impossible practically, we do not even know that universal analysis in the adolescent years would necessarily make for happier marriages.

We can only say that if we strive gradually to modify our whole educational process so as to give every human being from an early age a deeper insight into himself and his own needs; if self-knowledge in depth is no longer the forgotten factor of our whole educational system; if as a result of this we achieve earlier emotional maturity to keep pace with our intellectual and physical maturity; then in choosing our mates our conscious and unconscious goals are likely to coincide much more closely. When we achieve this, as part of many basic cultural changes and developments, then I think we can look forward to sounder choices and therefore to sounder marriages.

In other words, the capacity to choose wisely and soundly

and therefore to live in harmony depends upon developmental processes that must start in the early years, affecting the rate of maturation of the personality as a whole and the ultimate harmony between conscious and unconscious components in that personality. This is of far more importance in marriage than any kind of marriage counseling. The need for marriage counseling is in itself by implication a confession of cultural failure.

Moreover, by implication, a re-examination of the romantic tradition is indicated, and a careful and objective study of the obsessional quality of infatuation must be made. Most obsessions are, of course, quite unpleasant. Infatuations happen to be, in balance, pleasant obsessions, at least while they last. But an obsession is an obsession; and whether pleasant or unpleasant, it is never healthy nor conducive to ultimate happiness.

Granting that most marriage choices are made in the dark, as compromises between obscure conscious and unconscious purposes that are in conflict, what can one do about this? Does partial insight or any degree of retrospective insight help?

The answer is that insight may help in one of two ways: It may, and occasionally does, show that the marriage should never have been made. When this happens, two human beings instead of tearing each other to pieces may agree in all friendliness that each had made an innocent error of judgment in good faith. The tie can then be dissolved with a minimal amount of injury or pain.

In many other situations, however, a recognition of the unconscious as well as the conscious goals makes it possible for the married couple to help each other work out a harmonious compromise between their divergent purposes. Many a marriage has not only been saved, but actually has become deeply significant and constructive and harmonious through insight of this kind.

Has a marriage that is prearranged by parents, and in which the two young people take no part in the choosing, any advantages over our system?

We do not know that it has either advantages or disadvantages. It seems unlikely that the unconscious purposes of two sets of confused parents are likely to serve our aspirations any better than the unconscious purposes of a pair of confused youngsters.

Would trial marriages solve the problem?

Again, we cannot answer this. On theoretical grounds it is doubtful that trial marriages would be a solution. But since they have never been given widespread trial in an atmosphere that welcomes them, having always had to be conducted surreptitiously, we cannot say. We know that some people can learn from experience, and that others never are able to do this. This is true in every other aspect of life, and it is certainly true in marriage. We know people who, because of the shape and pattern of their unconscious problems, make the same mistakes in their choices over and over again. This can happen in trial marriages, just as well as in real marriages. Trial marriages hold no magic in meeting this problem.

How does it happen that some people manage to make happy marriages?

This is a searching question, and it is related directly to an equivalent issue: namely, how does it happen that some people manage to live not only without symptomatic neuroses but without even the burden of the masked neurotic disturbances of the so-called normal. We do not yet know enough to explain this fully. In medicine it is always true that we first begin to understand the normal by the study of illness. Thus the study of pathology throws light on normal physiology; and the study of neurotic deviations from normality is slowly building up an understanding of psychological normality. But we are not yet able to say *why* some few people develop normally whereas most of us develop neurotic episodes in early childhood, which leave scars throughout our lives.

In the present stage of our knowledge of human development those who escape this experience are happy accidents.

They hold up to us a shining hope. For if this can happen to a few people, we should be able to learn how to make it possible for most. When we have the answer to that problem, we shall know the answer to the problem of why, in spite of their lack of any deep personal insight, some people have, fortuitously and blindly, been able to make happy marriages. When we have this knowledge, a happy marriage will no longer be a happy accident. This, however, is going to take long and hard and serious study on the part of the human race. It will not be solved by clinging to old formulas which have not worked in the past.

Summary

We may best approach this problem through:

1. A modification of the romantic tradition, so that we cease to look upon this obsessional state as the highest artistic, aesthetic and spiritual experience of life.

2. A fundamental change in the processes of education, so that individuals grow up without the cleavage that occurs at present between the conscious and unconscious aspects of the personality.

3. The development of techniques by which such cleavages can be rapidly knit together as they occur in childhood, so as to re-establish the integrity and unity of the personality as a whole, especially as the young person approaches marriage.

You will see that I have no easy answers. I take the position that marital stress is nothing new in life; that it has been obscured by early deaths; that it is one of the oldest and one of the most important problems in human culture because of its influence on children as well as on grownups; and that it can be solved only by a frank facing of the fact that the capacity of one human being to love another is the most difficult and important challenge with which the human spirit is confronted, and that every effort of modern science must be brought to bear upon it.

It is a difficult and complex thing to be a human being, and the human race is not yet up to it. Part of the capacity to be a human being is the capacity of one human being to love another. This is a peculiarly human challenge. Let us not pretend that it is an easy one, to be solved by ancient formulae. On the other hand, let us not turn our backs on it because of the difficulties it presents.

References

1. Jacobson, Paul H., Vital Statistics Analyst of the Metropolitan Life Insurance Company, in the *American Sociological Review*, XV, No. 2, 235, 1950.

2. Jacobson, Paul H., personal communication, April, 1950:

"A precise measure of the extent of 'impulsive divorces among young people' would require a case history study of a scientifically selected sample of marriages dissolved by divorce. From the curves shown in Chart 1, however, it does not appear likely that such divorces are an important component of the total picture. The trend of divorces has been upward, even among families of long standing. In fact, the curves at the various durations of marriage are fairly parallel except during the late war and immediate postwar period, for those married less than 5 years. In other words, the sharp rise during the war resulted primarily from hasty war marriages, and also from the pushing ahead by several years of divorces which would have eventually occurred. In consequence, the recorded frequency of divorce was unduly high at the end of the war and in the first two postwar years; and this no doubt will be compensated for by rates below trend in 1948 through 1950 or 1951, after which the long-term upswing will again be apparent."

The Effects of Marital Conflict on Child Development

MARGARET S. MAHLER, M.D.*

RUTH RABINOVITCH, M.D.†

PSYCHOANALYSTS CONCEIVE of the family unit not merely as a group of individuals who deliberately choose to live together because of conscious affection and common interest, but, more than that, as a closely knit emotional organization of persons who have been attracted to each other and are held together by complementary unconscious motivations. The subtle immanent capacity of man to react to his fellow man's unconscious (empathy) and to seek out those toward whom he has, so to speak, unsaturated affinity (unconscious attraction) is especially important in the case of neurotic partners. The unsaturated affinity of neurotics is particularly strong, and they complement one another in their neuroses. In such cases of in-

---

* Clinical Associate Professor of Psychiatry, Albert Einstein College of Medicine; Member of the Faculty of the New York and Philadelphia Psychoanalytic Institutes.

† Formerly Psychiatric Consultant to the Jewish Board of Guardians, New York City.

trafamilial psychopathology, the family is held together by a pathological emotional balance which is easily disturbed.[3]

Though children's neuroses and maladjustments are due in large measure to intrapsychic conflict, both environmental and inherent factors play a role in varying proportions in the genesis of such disturbances. The process of growing up is in itself a very complex and difficult task. There are certain crucial points at different developmental phases which harbor particular hazards and stresses. Such stresses are increased by discord within the family. In a matrix of chronic marital discord, the parent objects offer distorted images. Their responses as emotional sounding boards to the child's reaching out for love cannot be true ones. Hence, the development of the child's object relationship is impaired and the task of directing and neutralizing the child's normal aggression cannot be achieved.

One notes a tendency on the part of child psychiatrists and other workers to assign the causes of emotional disturbances in children exclusively or predominantly to environmental factors. They believe and teach that all children would grow up happy and well adjusted if their environments were ideal. Clinical and statistical data, however, disprove this assumption. For instance, Dr. Louise Despert, who has made careful studies of extensive material, noted that the incidence of emotional disturbance among the children of divorce was proportionately less than that found in the general population.[2] While Despert feels that the emotional situation in the home is the determining factor in a child's development, we would prefer to state that it is *one* of the determining factors.

The children of incompatible parents show maladjustments which are in part reactive to and aggravated by the situation, but which are not caused by the marital discord. The conflicts which ensue in such children range through the entire gamut of the intrapsychic conflicts which we encounter in all disturbed children. In other words, it seems that marital discord itself does not produce any *specific* clinical picture.

It is true, on the other hand, that ever so often disturbed children are brought for analysis by parents who claim they have a harmonious relationship. Later we discover that discord actually existed. It is equally true that other children come for treatment because of severe neurotic difficulties, and yet we know that their parents' relationship is basically harmonious. In this latter group the child's conflicts are based on a combination of accidental and inherent factors whose interaction was unfavorable for the child's development. We should like to illustrate these comments with examples of children whom we have studied.

Gloria, age five, was brought for analysis because of an acute phobia of wild animals. She had a younger brother, age three and a half. Gloria's mother had been in analysis. Although the parents presented a façade of harmony, it seemed likely that this was not true. In the course of the child's analysis the dynamics of her phobia proved to be structurally similar to the classical childhood phobias described in the literature. A few specific features, however, are important.

Gloria had an uneventful childhood. There was no difficulty with her toilet training. She walked, talked and acquired her skills at the usual time. She presented finicky eating habits which became worse and took on the proportions of a serious feeding problem at the onset of her tiger phobia. Her mother noted that the child's agitation and fear were precipitated by her having been exposed to the hairy bodies of some men at the beach who had been enchanted by Gloria's unusual charm and graciousness.

During the early months of the analysis Gloria acted out in play her desire and conflict about showing herself and her curious looking at others. In her play she would set the table for the whole family. We then had to wait until Daddy came home. While doing so, Gloria would order the analyst to lie down and close her eyes, saying, "Don't dare to look!" During these games there was a marked contrast between her loving preparation of the meal for the homecoming father and the rest of the family, and unusual aggression towards the nurse-analyst and the baby-brother doll. The latter was often depicted as a little devil whose most important part was his large tail.

A period of wild, silly behavior at home and aggression against her brother ensued. Gloria often put a long object (such as a piece of cloth or a stick) between her legs while playing the devil and talking excitedly about her own tail. At this point her refusal to eat most foods was at its worst. Her rejection of her brother, a great clinging to her mother, and her curiosity about procreation all became manifest.

Her theories about birth were then revealed. She related with marked anxiety and excitement that her mother had been afraid she might get a ticket for parking at a bus stop during Gloria's analytic hour. In a thrilled and fearful manner she fantasied how Mummy night be put in jail and then Gloria would be afraid to go to the park because some strange woman might eat her up. We inferred that children are born as follows: Mummies (or women) eat up children whom they like and then the children are born to the Mummy who devoured the child.

Contact with Gloria's parents helped to complete our understanding of intrafamilial dynamics. The mother was a rather reserved and defensively sarcastic woman who had overly progressive ideas about rearing children and intellectualized her dealings with them. She expressed her love in such phrases as "I love you so much I could eat you up," but without any corresponding physical demonstrativeness. In contrast, the father's verbalizations were rather limited, but his behavior was overly demonstrative and overstimulating. When he left Gloria at kindergarten, he passionately embraced and pressed her little body to his chest in a prolonged parting ritual.

It appeared that the tiger phobia and its elaboration—the feeding difficulty—were over-determined. They resulted from Gloria's intense oedipal craving for her father which had been intensified by her experiences on the beach. Her phobia was also nourished by an equally adoring but entirely different type of person—her mother. Gloria's ambivalent clinging to her mother was determined both by her fear of being eaten by a woman and also by the wish to eliminate her mother, as in the jail fantasy. She counteracted this wish by clinging desperately to her mother and fearing to lose sight of her. This ambivalence was further reinforced by contributories such as aggression because Gloria felt her mother deprived her of the penis possessed by her brother. Yet since Gloria herself was a very passionate child, both in love and hate, the basic predisposition for her neurosis was in herself.

Analysis revealed that Gloria's anorexia developed as a reaction to the previously described birth fantasies and also as a reaction to the forbidden wish to introject "the little devil's tail." During a comparatively long period in her analysis, Gloria worked out her perplexity about hair—long or short, straight or curly, black, red or blond, and the location of hair on the bodies of men and animals. This was the immediate link to her masturbatory fantasies, fear of robbers and beasts, her dark-haired mother and her father's hairy chest.

When Gloria's phobia and anorexia had disappeared, her analysis had to be prematurely interrupted, although it was not yet completed. Her mother agreed to keep in touch with the analyst and return for consultation when necessary. Gloria entered school and became an excellent student and one of the most popular and beloved members of her class. Her neurosis remained evident, however, in the social area, where it would have been least expected. She felt constantly that no one liked her and that she was somehow different from other children. Although she accepted their advances graciously, she would never make overtures to her friends on her own initiative or arrange visits for them at her home.

During the course of the mother's analysis her reticence, though conspicuous, had to be respected. At this time the mother's analysis had ended, and we asked about the marital situation and how it appeared to the children. Despite her long acquaintance with analysis, her surprise at this question seemed genuine. It gradually emerged that little marital relationship existed and there was minimal mutual interchange as a family group. They had no recreations or interests in common.

When Gloria was about ten, the mother called with great urgency to request an interview for herself. During the interview she asked advice as to whether divorce or separation would be damaging to Gloria, since she could no longer endure the abuse and tension of living with her husband. We attempted to reassure her and asked to see Gloria again at this time in order to evaluate the possible consequences.

Gloria was happy to see her analyst again. She was obviously tense, but tried to present herself as carefree and nonchalant. She had been an exuberant, coquettish preschool child and a socially inhibited latency child. Now she had changed into an inhibited, introverted, somewhat depressed adolescent with massive repressions and a desperately maintained mechanism for denial. Gloria resumed her analysis for two years after her parents separated and her own neurosis was then worked through.

Gloria developed a phobia and accompanying symptomatology stemming from her oedipal conflict. This clinical picture is seen in countless cases in which marital discord does not exist. However, in the presence of such discord the child's symptomatology is intensified and aggravated.

In certain cases it is not immediately discernible whether a child's emotional disturbance became manifest because of marital discord or because of a certain constellation of inherent and early environmental factors.

Ludwig, the child of an ostensibly harmonious marriage, was brought for analysis at eight because of night terrors and panic reactions which occurred only at home. The family was a closely knit unit, dependent upon the fact that the father was a kindly, passive and motherly person who lavished all his affection and attention upon Ludwig. Ludwig's mother was ill and required bed rest after his birth, and the father assumed many maternal chores and ministrations. This reversal of parental roles continued and was further accentuated when Ludwig developed severe eczema, especially in the genital region, at nine months of age. Ludwig's father applied the medication and carried and soothed the baby when the pain and itching interrupted his sleep. Ludwig's parents consulted us when he was eight years old because his demands and attachment to his father threatened to disorganize the family unit and destroy this potentially harmonious marriage.

Ludwig was unable to sleep through the night alone. He slept in the parental bedroom until the age of five and recalls climbing over his crib or lowering the bar and tumbling into his father's bed. When he was five years old his younger brother was born and the family moved to a larger apartment where Ludwig and his baby brother shared one bedroom while his parents occupied the other. A tonsillectomy during this period increased Ludwig's fears. When he suffered acute panic and phobic reactions he could be quieted only by his father's taking him to bed. His father rationalized his acquiescence to the child's wish to sleep with him by using the time schedule of the ointment and compress applications as an excuse. When Ludwig started analysis he could only sleep by straddling his father's back or that of his mother when the father was absent.

Ludwig over-ate to the point of developing pseudo-Fröhlich feminine habitus. He further emphasized this by coy girlish mannerisms. He displayed petulant behavior toward his father and provocative aggression (often in the form of verbal obscenity) toward his mother.

During the course of his analysis he presented severe castration fears of many varieties, including fear of germs, appendectomies, robots, robbers and insanity. He was very preoccupied with thoughts of deformities in his own body and fears of bodily injury. Ludwig found the idea of a sexual relationship between his parents intolerable. His procreation fantasy was that the mother swallowed a pill and was cut open for delivery. Thus in his fantasies and in reality (through his nightly behavior) he succeeded in keeping his parents apart.

Ludwig's case also illustrates the use of the child as a buffer between the parents. Either parent may turn to a child for certain satisfactions not obtained from the partner. Such a misuse of a child's emotional dependency will further hamper his normal development. In Ludwig's case, his father took over the maternal role and gave the boy such exclusive attention and bodily closeness that the child's original sleep disturbance developed to the extent of requiring the physical presence and/or contact with either parent in order to sleep.

Aichhorn described this aptly when he said: "The love of a mother or a father which is meant for the adult partner and is lavished, in fact in a subtle way forced, upon a child affects his mental metabolism much as chronically indigestible, over-rich, unbalanced feeding would affect his stomach." [1] The intrapsychic conflicts, inevitable in the process of growing up, will become distorted and to a great extent insoluble without outside help. Without analysis, Ludwig, who is a basically well-endowed child, was headed for serious personality maladjustment.

Our third example is that of a child who also served a buffer role, but even more dramatically than Ludwig.

Five-year-old Nancy was the child of a discordant marriage. The acute symptoms for which she was brought to analysis occurred when she learned that her father was going to leave her (and "incidentally" her mother).

Nancy swallowed a brooch (which had been a gift to her from the departing father) as a symbolic way of introjecting his gift and simultaneously coercing him to stay with her by her suicidal attempt. The child was taken to the hospital immediately and the foreign body removed.

Subsequent analysis revealed an unusually overt oedipal situation between Nancy and her father, consisting of the father's sharing his bed with her, courting her and designating her as his "little sweetheart." There was little overt guilt on either side. Nancy's father showed his preference for Nancy as though she were an adult woman, and the mother again was relegated to a minor role.

Discussion

In cases of intrafamilial psychopathology the family is held together by a tenuous balance in which the child is made to play an important part. Whenever the relationship between the marital partners is not mutually satisfactory, the child's task of growing up will be additionally burdened. An anxious, unhappily married mother is a potentially poor partner in the first all-important relationship with the child. The situation is comparable to the "contagion" effects noted during the London blitz, when tense mothers communicated their anxiety to their children in such a way as to create an actual neurosis in them, whereas if the mother was relaxed the bombing and real dangers in themselves did not unduly disturb a small child.

It may appear paradoxical that the child's period of clinging to the mother may be prolonged in families with marital discord. This becomes understandable when we realize that the period of normal dependence (period of normal symbiosis) has been unsatisfactory and therefore has not freed the child to grow to the next stage of development, but rather has arrested his development in a fixation at this level. Clinically this disturbance will be apparent in prolongation of the pre-oedipal libidinal phases and related symptoms such as eating disturbances, enuresis, night terrors, etc.

Marital discord in the oedipal phase makes resolution of the conflict difficult, since this is accomplished through identification with the parent of the same sex. The small boy normally emerges from his oedipal conflict by identifying with the desexualized image of his father and by renouncing his aggressive and erotic claims toward his mother, replacing the latter with affection and tender feelings. Difficulties may arise in identifying with a father whom the mother dislikes or with a father who is experienced as hostile to his original love object, the mother.

The normal course of development is impeded if children are forced to play various unnatural roles within the intrafamilial psychopathology. One such role is that of a pawn, sometimes from birth on. Dr. Despert cites the situation in which a child is planned for and conceived to cement a marriage which already stands in jeopardy.[2] These children, from their earliest life have been given an enormous task which their parents themselves could not achieve. The unrealistic and childlike expectations of parents in these situations are only too clear. Such children tend to suffer from great emotional insecurity, wavering loyalties, ambivalence and guilt feelings which may lead to ego and super-ego defects, inability to form stable object relationships, and may result in character deformities, depression or other narcissistic neuroses.

Closely allied in its consequences to role of pawn is the role of confidant to either or both parents (simultaneously or alternately) who are hostile to each other. As the confidant, the child is burdened beyond the capacity of the immature stage of his ego development. These parents rarely spare the child such traumatic and perplexing details as those pertaining to cruelties inflicted by the partner or luridly suggestive material pertaining to their sexual experience. The child is subsequently hampered in his psychosexual development, and either turns away from sex altogether or else retains immature emotional attitudes and distorted patterns instead of normal heterosexual relationships.

Children are forced into these roles because of the individual neurotic needs of their parents. It is our impression that a critical factor here is the degree and quality of the adult empathy for the child *as a child*. This question has been discussed by Christine Olden, who clearly distinguishes various adult emotional attitudes toward children from mature empathy.[4] It would be fallacious to assume that the adult neuroses which find expression in discordant marriages are accompanied by identical empathy defects. The individual variation of empathy defect is considerable, but the existence of such defect can be inferred from cases in which children are forced into the unnatural roles of buffer, pawn and confidant. The question of the child's emotional experience with other significant adults and their capacity for empathy is important. Sometimes such substitute relationships afford an escape from severe pathology to a constitutionally strong ego.

Children may be injured by impulsive actions on the part of marital partners, such as abruptly confronting the child with major decisions or changes. Here again is the adult expectation that the child adapt himself promptly and without preparation to radical changes in the environment. Parents who make these demands because of their own needs are unaware of the difficulties involved for their child.

Equally damaging are lengthy, recurrent or violent scenes. Such scenes may alternate with reconciliations, or discord in which tension and distressing falsehood replace overt aggression. Children who live through these experiences feel even more perplexed, threatened and lonely.

Most pathogenic to the child's development is protracted subtle discord. Despert designates one variety of this chronically pathogenic situation as "the predivorce uncertainty." [2] The child cannot cope with such a conflict situation because it is not out in the open. Often he is even expected not to notice it. This, in turn, reinforces the pathologic defense mechanism of denial.

In cases of actual separation or divorce, resulting in the absence of one parent, the child may develop one of two typical tendencies. Either he idealizes the absent, withdrawn or little-known parent, or he tends to devalue him completely. Devaluation may help him retain a feeling of loyalty toward the remaining parent and earn the good graces of the one upon whom he is now increasingly dependent. Devaluation may also be used to ward off the hurt inflicted upon the child by the absent parent's disinterest or desertion.

Obviously, with either alternative the processes of identification and organization of an integrated super-ego will be rendered difficult, if not impossible. In our observation of such children at different ages (preschool, latency and adolescence), we often note a tendency to pay lip service to one of the parents. The child seems to accept and repeat the verbalizations of the more hostile parent. Actually, he is thrown into deep inner conflict which becomes apparent as he gets older. His unbalanced identifications and disturbance in the solution of normal developmental conflicts may be expressed in neurotic symptom formation as well as unpredictable acting out and finally in serious deformities of character. During early life the human being is destined to be dependent on his parents, and later on the images he has erected within his own self of the two parental figures. Sometimes a disgruntled and defensive youngster is secretly mourning the lost, disparaged parent.

Considerations in Therapeutic Work

Today there is an increasing tendency on the part of such parents to seek professional advice and assistance. That there are no simple solutions is clear. It is important to understand each member of the pathologic family constellation—an understanding that presupposes a high capacity for empathy with each member and an ability to remain objective. Only by empathy with each member can the counselor gauge the emotional capacities and limitations. He must work within these limitations, knowing that no efforts of will and intelligence alone on the part of a human being will enable him to transcend these limitations.

Thus an hysterical or impulsive mother who has been asked to avoid making scenes which are damaging to her child will experience increased tension and guilt feelings without being able basically to alter her behavior patterns. A father who is engaged in an extramarital affair, with ensuing tension between him and his wife, will not be able to spend more time at home with his children even though he loves them dearly. Even when one has obtained the best possible insight into such a family situation, one must still refer to the specifics of each case before deciding whether it is better for the parents to divorce or remain together "for the sake of the children."

While the imprints of marital discord do not always lead to manifest neurotic symptoms in the child, they affect his attitudes and outlook on life. Later these appear as unconscious patterns which impair his own choice of a sexual and marital partner. The child now grown to adulthood may repeat in a similar or complementary way traumatic situations which the marital discord of his parents stamped on his pliable personality structure as a child.

References

1. Aichhorn, August, personal communication.
2. Despert, J. Louise, *Children of Divorce*. New York, Doubleday & Co., Inc., 1953.
3. Mahler, Margaret S., "Freud's Viewpoint for Child Guidance." *Handbook of Child Guidance*. Ernest Harms, ed. 685-706. New York, Child Care Publications, 1947.
4. Olden, Christine, "On Adult Empathy with Children." *Psychoanalytic Study of the Child*, Vol. VIII. New York, International Universities Press, 1953.

Neurotic Choice of Mate

LUDWIG EIDELBERG, M.D.*

TO MOST ANALYSTS the concept "neurotic choice of mate" connotes a marital selection which in itself interferes with normal relationships or makes them so difficult that the displeasure exceeds the pleasure derived therefrom. Whenever such an error in judgment has been made, we may justifiably conjecture as to whether or not it was caused by the patient's neurosis. One must then proceed by proper analysis to build a chain of evidence that will establish beyond doubt the interference of the unconscious.

A neurotic choice of mate may be the result of various unconscious defense mechanisms. Fixation or regression to one of the three stages of development generally lies at the core of the problem. Thus, defense mechanisms that are directed against an awareness of phallic wishes may be separated from defense mechanisms originating at the anal or oral stages. All three

---

* Associate Clinical Professor of Psychiatry, State University Medical Center, New York.

types of defense mechanisms can interfere with the choice of
mate by making a person select someone who helps partly to
gratify and partly to deny the presence of infantile wishes. In
other words, whenever a neurotic choice is made, the patient,
instead of choosing a person with whom he could be happy, has
selected an object he needs in order to avoid recognizing what
he is afraid of. The defense mechanisms used to achieve this
aim lead to various pathological formations that are ego accepted and can therefore be differentiated from neurotic symptoms.[1]

There are many ways to classify the variations involved in a
normal choice of love object. Freud suggested a division into
two groups: in selecting a mate we may either look for someone
who is similar to us, or become attached to a person who would
protect us. As an example, Freud pointed out that under normal
conditions a little boy will select his father as the object he
wants to be similar to, and his mother as the object he wants to
be nursed by. Consequently, we distinguish between a narcissistic object choice—an object similar to us (someone I want to
become similar to, or someone I want to make similar to me)—
and an anaclitic form of object choice—an object we need to give
us what we don't have (food and protection).[3]

However, a normal boy will not only regard his father as the
person he wants to be similar to, but also as the one who gives
him the protection and the strength he himself lacks. At the
same time, his mother is important not only because she nour-
ishes him, but also because she serves as an object he tries to
imitate.

The narcissistic factor is generally considered the more important one in the relationship between a boy and his father,
whereas the anaclitic form of choice dominates the relationship
between son and mother. The opposite is true for a little girl in
relation to her parents.

Normally, an adult male in search of a mate will be more interested in finding a woman who represents chiefly an anaclitic choice of object. While such a woman will undoubtedly gratify some of his narcissistic needs by her interest in his problems, he will not eliminate women who are dissimilar to him, nor will he try to become similar to them or make them similar to himself. The normal man will gratify most of his narcissistic needs by friendly relations with other men. Something comparable appears to take place in the object choice of a normal woman.

When an individual is interested primarily in the gratification of his narcissistic needs by a heterosexual object, or uses an anaclitic approach in his selection of objects belonging to the same sex, a danger of neurotic conflict exists. An overt homosexual, for instance, is unable to accept as his mate an object belonging to the other sex, and therefore different from him. He may accept an individual belonging to the opposite sex as a friend only so long as such a friendship remains free of genital desires.

The following brief examples may serve to illustrate some of the problems encountered in the study of neurotic conflicts in marriage, involving the choice of love object. It should be noted that while the patient suffering from a symptom is aware that his suffering is caused by his illness and is eager to be treated and cured, the character neurotic is not aware that he is sick. He consults an analyst because he is dissatisfied and unhappy with the pattern of his life. Often, an unhappy marriage is a means whereby the character neurotic expresses and tries to gratify what his unconscious wants, but his conscious personality resents.

Carol came for treatment after her third divorce. She offered many excuses for her three failures, but only analytical investigation was able to uncover the real cause of her troubles: Her infantile wish to marry her father, of which she had been unaware, transformed any man she was married to into an incestuous object. No sooner did she marry a man with whom she thought she could be happy than he began to mobilize in her feelings of guilt. Unwilling to face the disturbing element, the patient then had to search for another man who would take her away from her husband-father. Hence her repeated errors in choosing a mate.

David was married to a cold and unattractive woman, ten years his senior. While he complained about the choice he had made, he hesitated to divorce the woman whom he disliked intensely. Analysis disclosed that his frigid wife was the only woman with whom he was potent. Before and during his marriage he had tried to have sexual intercourse with women he found desirable, but failed to have an erection. His passive feminine desires interfered with his masculine role. When he was attracted to a pretty woman, he not only wanted to have her but also to be like her. This wish to have a vagina in addition to a penis made the execution of the sexual act impossible.

The patient was potent with his homely wife because he had no wish to identify himself with her. While he consciously resented being tied to a woman so different from the ones he pictured in his daydreams, he was reluctant to leave her because it meant depriving himself of the only outlet for his genital desires. Confronted by the choice of being impotent or unhappily married, he preferred to remain with his unattractive and aggressive spouse who allowed him to discharge his hostility and remain free of guilt feelings.

Ann was a plain-looking woman who complained about her husband's infidelity, but was unwilling to divorce him. Her husband had many affairs and took no pains to keep them secret from her. The patient felt that she had no other recourse but to accept these humiliations because all men were polygamists. However, a study of her masturbation fantasy disclosed that she could be genitally excited only by a sadistic partner. Moreover, it turned out that by an unconscious identification with her husband, she was able to play the role of the seducer and in that way gratify her unconscious homosexual tendencies.

The dynamics of the treatment of Mrs. S. are presented in more detail in order to illustrate the many complicated factors involved in a neurotic choice of mate.

Mrs. S. began analysis, after she had been married a short time, because of her frigidity. At the beginning of her married life she had experienced some degree of climax, but gradually her ability to obtain pleasure from sexual intercourse decreased and then disappeared. When she first came to my office she regarded sexual intercourse only with disgust. She believed that marriage had destroyed the elements of romance and love she had felt for her husband while he was her lover, and blamed her frigidity on the drabness of her marital situation.

It cannot be denied that extramarital relations may increase the sensation of pleasure by adding the element of aggression which is discharged by doing something forbidden.[2] Also, it is true that a person may become more aroused initially when sexual relations are surrounded by an aura of surprise and glamour. Mrs. S. felt she could easily restore her lost genital sensation if she were unfaithful to her husband. Yet she considered herself too decent to go to bed with another man. She was therefore resigned to sacrificing her sexual pleasure on the altar of matrimony.

The study of masturbation and of the masturbation fantasy can often increase insight into the structure of the neurosis. The manifest content of the erotic fantasy, different from the manifest content of the dream, is always ego-accepted, and can therefore be used to demonstrate to the patient what it is he actually wants. In the course of treatment the patient revealed the fact that she was still having genital sensations and a clitoral climax while masturbating. Mrs. S. fancied herself being undressed, caressed, and finally having sexual intercourse with a charming man. From a superficial point of view, her masturbation fantasy did not show the presence of an obvious pathology. She explained the paradox that masturbation provided genital stimulation, while sexual intercourse left her cold, by her ability to be promiscuous in her daydreams. "If only I had not been so faithful to my husband," she said, "I could have achieved what I have been missing."

A deeper study of her masturbation exposed some significant factors. While the pleasure she obtained was caused by the promiscuous variety of objects she used, she could neither describe them nor say whether they reminded her of men she knew or had seen in the movies, the theater, etc. However, she could sense their excitement when they looked at her and their pleasure in caressing her. Furthermore, although she had no optical image whatsoever of these men, she saw herself very distinctly in her fantasy. Subsequent insight revealed that the image of the woman in the fantasy was not that of the patient but of her mother.

Occasionally a patient develops not only a positive, but also a negative Oedipus complex, and his neurosis represents an attempt to cope with both problems. While it appeared that at least one of her unconscious problems was due to a negative complex and a consequent identification with her father, Mrs. S. had relatively little resistance to accepting my suggestion that she wanted to play the role of her mother and sleep with her father. Nor did this interpretation contradict the strong identification with her father.

The fact that the patient could feel what the man in her fantasy was supposed to be feeling, without seeing his image, while she could describe fully the female image suggested that she identified herself with the man in her fantasy. Despite strong resistance to this idea, Mrs. S. admitted that she did not share the feelings of the woman used in the fantasy, but enjoyed her excitement as if she were watching and listening to her. Mrs. S. finally recognized that while she seemed to be contemplating infidelity, she was unconsciously using her promiscuous wishes to ward off her latent homosexual desires.

In short, understanding of the masturbation fantasy in connection with this patient's analysis was most helpful in locating the unconscious elements responsible for her choice of husband. Mrs. S. was able to recognize the fact that in the past she had avoided men who were in love with her because they could not conform to her concept of what she expected from men. Her husband had been accepted because he seemed to be a weak individual, overwhelmed by his love for her. She identified herself with him, and in this identification she loved herself as if she were he. She had selected her husband not because she wanted to have him but because she hoped to use him to obtain what was not available to her.

Often the recognition of unconscious factors responsible for the choice of the love object leads to the rejection of the person originally chosen. In this case, however, it became apparent during the analysis that the patient's husband was not as weak as she had thought him to be. After her neurotic blind spots were removed, she was able to see her husband in a new light and fall in love with him in a healthy way. The marriage was salvaged.

The normal individual is not infallible, and is bound to commit some errors. These errors, because they are not caused by unconscious factors, will be recognized and corrected in due time. The neurotic, on the other hand, is either unaware of the error when it is committed, or is unable to avoid making it because it represents a valuable outlet for his neurotic tension.

A normal person may find himself attached to the wrong mate as a result of a mistake in judgment. A neurotic, who uses a neurotic approach in selecting his love object, will consistently manage to seek out someone who must produce trouble because he is able to serve as part of a defense mechanism. Without analytic understanding the patient will be successfully fooled by the hidden infantile wish, and will be forced to repeat his mistake.

The problem of neurotic choice of mate, like other neurotic problems, can be solved only by analyzing the patient—not his marital problems. Only after an analysis has been terminated successfully can we attempt to isolate a certain problem for didactic purposes. Even if we had the ability, which we do not, to eliminate a patient's neurotic choice without dealing with his other neurotic defense mechanisms, the practical results of such an isolated therapeutic act would be nil. The patient, deprived of the weapon used for the discharge of his repressed wishes, would employ other defense mechanisms or create new ones to replace what was lost.

While we know enough to help a patient who cooperates to free himself of his neurosis, we know very little about the causes responsible for what is usually referred to as "the problem of the choice of the neurosis." Almost nothing has been published that answers the question: Why does one patient use symptoms whereas others develop neurotic character traits or perversions to ward off infantile wishes? In general, however, the developmental factors which are responsible for the choice of the neurosis are also responsible for the choice of a mate.

It is obvious that the nature of the marital discord varies with the type of neurotic choice of mate. The problems arising from an unconscious sado-masochistic choice, for example, will be quite different from those rooted primarily in the compulsive need to take a man away from another woman. However, many elements of neurotic choice may co-exist, although one predominates. Analysis regularly exposes the network of neurotic interaction caused by unhappy choices. The therapeutic task in such cases is to attempt to establish a better equilibrium in the interaction between partners, if that is at all possible.

References

1. Eidelberg, L., *An Outline of the Comparative Pathology of the Neurosis.* New York, International Universities Press, Inc., 1954.
2. ———, *Studies in Psychoanalysis.* New York, International Universities Press, Inc., 1951.
3. Freud, S., "Introduction to Narcissism." *Collected Papers,* Vol. IV. London, Hogarth Press, 1934.

The Unconscious Meaning of the Marital Bond

MARTIN H. STEIN, M.D.*

THE SATISFACTORY ANALYSIS of a married patient requires the most thorough examination of all of the unconscious meanings of the marriage itself. Omission of any major portion of such material may either lead to the use of the marriage as a perpetual resistance, resulting in a stalemated analysis; or, on the other hand, impulsive acting out in the direction of divorce and remarriage.

For the purposes of this study, I have chosen to discuss a single major determinant, an unconscious fantasy concerning the nature of the marital bond, which was the expression of serious distortions in the attitudes of four neurotic men toward themselves and their wives, and one which played a prominent role in the entering into and perpetuation of their marriages. In

---

* Lecturer, New York Psychoanalytic Institute.

its simplest form, the fantasy may be expressed as "My wife is my phallus" or "My wife is part of my body."

In each of these patients this fantasy played a significant, but by no means exclusive role in their serious marital problems. Its clinical importance lay in its unconscious anatomical significance, which had to be understood clearly and worked through by the patient before much progress could be made in solving problems in his marriage—or, for that matter, in other aspects of his life situation.

The first patient was an intelligent young man who had a responsible job in an advertising firm and was considered successful and happy by most people. He married a girl he had known for some years, whose background was very similar to his own—well educated, attractive and competent. This was a marriage of which both families approved. It is true that he went through a period of considerable anxiety and doubt before he married, but he has been a faithful husband and good provider. Their children are healthy and attractive. Their many friends considered them a good example of prosperous, well-adjusted and intelligent "young marrieds."

This idyllic pattern does not stand more than a very superficial investigation, however. Actually, the patient's married life had been very unhappy. He was afraid of his wife, and would often neglect his work because of his fear of her displeasure. She would scold him bitterly, leading to scenes which culminated in his weeping. Only his own competence and his employer's tolerance allowed him to hold onto his job without excessive difficulty. He had very marked indigestion and suffered from many colds. Frequent sore throats caused him to miss work for several days at a time. He visited his family physician constantly, pleading for some new medicine that would cure these illnesses.

Outwardly, he was a long-suffering, severely henpecked husband, constantly manipulated by his wife and always resentful and frightened. His conscious fantasy life, however, was quite different. It consisted of daydreams of sleeping with tall, beautiful, blonde prostitutes; of making his wife a slave and using her for his own sadistic purposes, beating her and forcing her to masturbate him.

To his wife, meanwhile, he was a suffering and plaintive husband, often ill and very passive in most of his attitudes, weeping readily when she scolded him, and not very competent sexually.

Only the recognition of his masochism, as a reaction to his repudiated sadistic fantasies, was required to bring about alleviation of many of his more obvious difficulties. He stopped weeping, talked back to his wife, and was often able to remain calm in the face of her increasingly provoking attacks. His somatic symptoms were greatly improved. He then became aware of more painful and deeply concealed fantasies about his marriage.

As a bachelor, his fondest wish was to have a tall, beautiful, blonde girl, who would appear with him at dances, drawing admiration to him, and who would later indulge in mutual masturbation with him—he was not particularly interested in intercourse. He had been surprisingly successful in his efforts to attract girls. To none of them, however, did he develop any lasting attachment. Mature girls, with backgrounds similar to his own, were not attractive to him, but finally, with great anxiety, after a stormy engagement, he married a small brunette of his own social group. She was competent and vivacious, but inwardly unhappy and frightened. The first year of their marriage went fairly smoothly, but after a year, the birth of their son was followed by great anxiety in both of them, and there began a series of bitter sado-masochistic quarrels, which continued up to the beginning of the analysis.

It became evident that, although he had hoped consciously for a rather boyish, competent wife, who would reassure him about his own fantasied deficiencies, he wished also for a wife whom he could manipulate and who would be his slave. His wife would accuse him of using her for his own purposes without loving her, as if she were a prostitute, and of treating her like a mere tool. There was, as we shall see, much justification for her complaints. He often fantasied that she existed only for his pleasure, to masturbate him, to bear his children and cook his dinner, and be beaten by him when he felt like it. Meanwhile, his outward behavior was that of a veritable Milquetoast.

Gradually, we became aware of the meaning of the sadistic fantasy which he dared not act out. As a child, he had been reared by anxious and quarreling parents and, for this and associated reasons, became more than usually timid, retaining a strong feeling that he was not as masculine as other boys, that his penis was smaller and that he would never be a powerful, sexually potent man. As a child, he insisted openly that this was due to his mother's failure to feed him properly. A good deal of sexual play occurred with his little sister, whom he regarded with mixed contempt and affection. Later, he developed the idea that she would always be his responsibility, that she would never marry and would be a "burden" to him. When, recently, she did marry, he reacted as if he had experienced a violent loss.

In adolescence, he reassured himself by his popularity, particularly his ability to attract tall, pretty, blonde girls, but his feeling that he was castrated never left him. In adult life, this castration fantasy was manifested by annoying pains in the testicles and a fear that his penis was pathologically small. He was unable to urinate in front of other people and his potency was always impaired. His feelings of inadequacy were reflected in his attitude toward his work and toward other men, particularly large, bluff "masculine" men, with whom he felt very inferior. Only when he had a girl with him, for example, a pretty secretary, did he feel reassured and somehow a man among men. Even the shortest separations from his wife resulted in great anxiety.

The meaning of this became evident through a long series of associations and dreams. It was illustrated most dramatically in one dream, which I report fully.

"I was standing on a mountain or cliff. A big man, standing to the left and behind me, threw me a football which I caught between my legs. I fumbled it for a moment, but finally clutched it to my chest. Then my father threw me a basketball. Both balls seemed to be dropping into the valley very fast, toward some sort of stream—I wasn't too worried about it."

His associations were as follows: That night he had returned from a business trip to find his wife extremely affectionate. They had intercourse by having her sit on his lap and then lie on him, while he clutched her to his chest, as he had clutched the football between his legs and then to his chest in the dream. Before penetration, he had excited her manually. This was represented in the dream by the fumbling of the football. Coitus was very enjoyable and mutual orgasm occurred. The tall man of the dream was in the position that I occupy during his analytic sessions.

He thought about his mother and father. They had both condemned masturbation very severely. Recently, he had been feeling quite differently about it. He had sensed that my attitude was not punitive and that I would not punish him for masturbating as his parents would. The football, he thought, was pointed like a penis, or like the testicles, while the basketball is round, without a point. I asked what other differences there were. He replied, "You may hold a football tight, play with it, and kick it around. You may only dribble a basketball." It struck him that this was the only way he was permitted to use his penis as a child—that is, for dribbling or urination.

He commented that his marriage seemed much better now than it was. He hadn't managed it well before, but now he was more confident and firmer. He could handle things well, even his difficult wife. During intercourse last night, having his wife sit on him gave him a good feeling—it was as if she were an extension of his penis.

This is an abbreviated account of his associations, but I have included those most clearly linked to the material of his dream. On one level, the dream may be interpreted as follows: He is now permitted to keep his wife close to him and he is no longer a fumbling husband. Unlike his prudish parents, I have permitted him to play with his wife and to have sexual relations with her without fear.

This, of course, is only part of the story. In the dream he expressed one of the most important unconscious fantasies about his wife. He feels more of a man, less threatened by his punitive father, who had permitted him only to urinate or dribble with his penis; now he may play with it, since he does not have to be afraid of me. Me may have a penis, plaything, toy, or wife, whom he can hug to himself, fumble between his legs, and (remembering his sadistic fantasies) kick around. He fantasies, therefore, that it is his wife who is his penis, sitting erect on his legs. He excites or plays with it or her until he produces an ejaculation in his penis or orgasm in his wife. This deeply-buried fantasy, derived from his childhood, had always been a powerful factor in this adult man's relationship with his wife.

It is possible now to explain some of the aspects of this patient's marital problems which were so obscure before. He really did, in a subtle and unaware fashion, treat his wife like a "tool," he played with her, was quite unable to regard her as a human being, but rather as an appendage, a part of his body.

Unconsciously, his sexual relations with her were a kind of masturbation, independent of her real needs and feelings. Why had this fantasy remained so important?

Castrated, as he imagined himself to be, he always felt that he was in need of a penis. His active, competent, scolding and vivacious wife, by her attachment to him, repaired or undid this deficiency and, in his fantasy, became his phallus.

This had several effects, other than those mentioned. In spite of his unhappy life with her for a number of years, he never for a moment considered separation or divorce seriously, for how could a man separate himself from his phallus? He would be no man at all without her. Even when his fantasies did hint at separation, he had to imagine a taller, blonder, prettier girl waiting for him.

Of course, his sadistic fantasies did not merely represent hostility to his wife, which was not so very extreme; they also expressed his need to excite and manipulate his wife, to masturbate by beating the penis-wife, to cause her to weep, that is, to urinate or ejaculate. He could not think of her as a human being. She sensed this, and reacted accordingly, once saying quite spontaneously in a fit of anger, "You don't want a wife, you want a penis and balls to play with."

Finally, even the unconscious infantile fantasy (my wife is my phallus) may be used in the service of clinical improvement. Now she is something he may approach and handle without fear. The dream was a primitive and regressive expression of the wish to maintain what was actually an improving relationship.

The other patients to be presented demonstrated the same fantasy, operating in somewhat different fashion.

The second patient was a successful, intelligent and hard working broker in his thirties, who came from a family which had some social aspirations, hampered by an uncertain income. One of his motives for marrying his wife was the attainment of this social position and economic security which had always been just out of reach. They had much in common, in background and education. They have been married a long time and have several children. No bitter quarrels have occurred.

Nevertheless, there has been very pronounced unhappiness. Often he thought of dissolving his marriage, but he wondered whether he could have been successful without his wife's position and money. (Actually, there was little reality in this. He was a man of great ability, and would probably have done well in his business without either so much money or social position, although he might not have been able to live so extravagantly.) His relationship with his wife became more and more attenuated and he had a number of abortive affairs with other women, with some of whom he thought himself infatuated. When he thought of divorce, however, he did so in carefully guarded ways, which revealed his great fear of such a step. He felt his business would fall apart, that he couldn't support a new family, or that his wife would commit suicide as soon as they were separated. He brought out nothing to support these fantasies, except for his feeling that she couldn't survive without him because of her helplessness. He, in turn, doubted his ability to support a family without her wealth.

It is true that as the marriage went on, she became more helpless, while he took over many of the duties he felt were hers by right. Outwardly he expressed great annoyance about this, but one of his early dreams revealed his true attitude. He dreamed that he had a collection of wooden figures of women, all arranged in shelves. His associations had to do with his attitude toward his wife and toward other women—women were really "objects" to be collected. An interesting technical point in this connection was his inability to give me any clear description of his wife, or, for that matter, of any other woman. His associations about them were completely colorless and lacked feeling, although otherwise he was articulate and well read.

At one point in the analysis, he had a dream which dealt with confused memories of his parents fighting and having sexual relations. This was discussed but not interpreted beyond pointing out his sado-masochistic impression of his parents' marriage. The next day, he reported the following dream:

"I saw two men, and a crowd watching them. There was some sort of horseplay going on. The man in the rear stuck the front one in the buttocks with a stick; then he would take the stick away and

wave it in front of the crowd to show that it was covered with something like tar. Everyone laughed. Then I was talking to a customer
who told me that my prices were too high. Finally, I was walking
down a stairway very slowly; it was quite slippery and wet; my
wife was in front of me and she fell backwards, her body rigidly
straight. She seemed to fall very slowly in front of me."

His associations were as follows: This dream was reported on a
day on which he brought me a check for my fee. He had been carrying the check around for a full day, but had forgotten to give it to
me. The high price mentioned in the dream was one that he was
charging someone for some goods. The price was not really too high;
it was my fee he regarded as exorbitant. He was very much concerned about money and his standing in the business world, whether
he could make a living by himself, whether he was dependent upon
his wife's family. Although he did not really love her, if he should
divorce her, he would be unable to support himself and a new wife.

In being charged a fee, he felt shamefully treated, like the man
who had the stick pushed into his anus. When the stick was removed
with the feces (unconsciously equated with money)—this was being
treated like a woman, being raped and deprived of his strength and
dignity.

He sought a remedy for castration. In the stairway episode, his
wife had become his penis, sliding along the slippery canal. He did
have a big penis after all—a wife. He was using her as he had used
his younger sister in childhood. He had been a very timorous boy,
unable to admit it. Alone in the house with his sister, he would become very apprehensive. He would then convince the little girl that
she was the frightened one, and telephone his parents to say, "Come
home, Annie is scared to be alone."

In addition, there had been sexual play with his sister in which
he was the aggressor, feeling all the while that he was unable to stop,
as if he were masturbating. In early adolescence, this activity had
in fact been a substitute for the masturbation one would ordinarily
expect.

When later, his sister became dependent on him for social contacts, he felt contempt for her. She was regarded by him as a de-
spised, yet necessary appendage, a scapegoat-penis. These attitudes
were transferred almost unchanged to his wife, whom he regarded
as helpless and inferior, yet somehow necessary to his potency.

Thus, the wife-penis fantasy in this case is very similar in origin to that of the first patient presented. The marital situation is obviously different, however.

I shall describe the third and fourth patients briefly.

The third patient, a man with a successful career, was beset by many feelings of inadequacy and by homosexual fantasies. His pretty wife was much younger than he. When they were married, she was a helpless and frigid girl with disabling hysterical symptoms, which caused the patient much concern, often interfering with his work. She was psychoanalyzed and improved a great deal, however, becoming a relatively self-sufficient, even self-assertive young woman, no longer so helpless and dependent on her husband, and for the first time openly eager for sexual satisfaction. This threw the patient into a severe panic, manifested chiefly by increased difficulties in coitus.

During his analysis, he had the following dream:

"I was walking with a friend. My wife was walking a little ahead of me, just window-shopping."

The following were his associations: The window-shopping referred to shops along the main streets of many beautiful cities the patient and his wife had visited. These shops were full of beautiful things that they couldn't always afford. The night of the dream he had wakened from sleep with an erection, but, for some reason, couldn't bring himself to awaken his wife and attempt intercourse with her. He was really only "window-shopping" with his penis.

In the dream, his newly-cured and strengthened wife was his penis, as well as the object of his sexual attraction. The wife-penis again undid the castration he felt so keenly in his fantasies, which were marked by depressing thoughts of poverty and illness.

This interpretation was confirmed a few days later when he met the attractive fiancée of a powerful politician. As he shook hands with her, he became very anxious: "It was a homosexual feeling." That is, touching her hand represented touching the powerful man's most treasured possession—his penis. For this patient, therefore, the great man's fiancée was his penis, as his own wife was to him. He loved to fondle his young wife, and was really attached to her, but he did share with the other patients reported here, the tendency to treat her as a possession and to resent any real independence on her part. There were also sexually charged fantasies of choking or squeezing his wife, again as if she were a penis to be masturbated.

On one occasion, while his wife was away on a visit, he was invited to dinner alone. He felt forced to refuse. "I was afraid to go alone. I don't know what it could be; it was as if I couldn't meet people without her. She's my front."

The fourth patient who demonstrated this fantasy had the following dream during a period of concern over his potency:

"I was standing up and walking around. Somehow I was carrying my wife on my hips, as if I were having intercourse with her in that position. She was sitting straight up. I felt very strong and elated."

This couple used to act out a little game in which the husband would fondle his wife's head as she sat on the toilet. This would induce her to urinate, thus again making her his penis, with which he made water. His attitude toward her was not unlike that of the other patients reported; he regarded her as a beautiful possession, to be shown off, something with no identity of her own. There were many sado-masochistic features as well. Discussion of divorce would produce the most troublesome fantasies of being castrated, for example, of contracting mumps and becoming sterile. In both this patient and the previous one, rescue fantasies involving the wife were very prominent, accomplishing not only the rescue of the self, but also of one's own penis, which was felt to be in such great danger.

The unconscious "girl equals phallus" equation (or "The woman is an appendage of my body"), is one which has been familiar to analysts through the work of Ferenczi,[2] Lewin[3] and Fenichel.[1] Its converse, the unconscious fantasy of the woman that she is an appendage of the man, has been vividly described by Annie Reich.[5, 6] Whether women who fantasy themselves as appendages marry men such as I have described is a question which cannot be answered categorically. It was clearly expressed in the first case reported here, but was less clear in the others.

One of the most perceptive and entertaining accounts of the phallus-girl equation is found in George Bernard Shaw's "Pygmalion." Professor Higgins, a devoted student of linguistics and a confirmed bachelor of somewhat eccentric tastes, has, for clearly selfish reasons, transformed a Cockney flower girl, "a squashed cabbage leaf," into a beautiful and charming creature, easily mistaken for a duchess. This miracle has been accomplished chiefly through the correction of her speech.

In company with an elderly bachelor colleague, he exhibits Eliza, his Galatea, at a tea party in his mother's home. His mother, a most intelligent lady, reproaches them, "You are certainly a pretty pair of babies, playing with your live doll." Later, when Eliza attacks him for his lack of human feeling for her, he defends himself, "I care for life, for humanity; and you are a part of it that has come my way and been built into my house." Finally, Eliza, basically a very tough-minded daughter of a Shavian-realist dustman, defies Higgins and attacks him for his attempt to use her as "a baby or a puppy"— or a slave. She asserts her independence, and after Higgins experiences a period of great anger and panic, he pleads with her: "Five minutes ago you were like a millstone around my neck. Now you're a tower of strength, a consort battleship. You and I and Pickering will be three old bachelors together instead of only two men and a silly girl."

Shaw had an excellent sense of reality, and did not, in the epilogue to his play, allow Eliza to marry Higgins, who was unable to regard her as anything but an appendage to be used and exhibited.

"Now, though Eliza was incapable of explaining to herself Higgins' formidable powers of resistance to the charm that prostrated Freddy at the first glance, she was instinctively aware that she could never obtain a complete grip of him, or come between him and his mother (the first necessity of the married woman)." As usual, Shaw has struck at the heart of the argument. He ends the epilogue, "Galatea never does like Pygmalion: his relation to her is too godlike to be altogether agreeable."

The movie makers were more romantic. They had Eliza

marry Higgins. You may speculate for yourselves as to whether the marriage would be a happy one.

The vicissitudes of a fantasy such as this may be studied with reference to two opposite directions: regressive, toward its origins in earlier phases of development; and forward, toward its appearance in the symptoms and character traits of the adult. The latter is far easier. The manifestations of the fantasy, as it appears in each of the four men I have described, are fairly clear, and it is not too difficult to demonstrate its specificity. In these patients, the girl-phallus equation serves to bind the marriage; it gives greater force to the concept of the adhesive character of marriage, so often expressed in the marriage service, in homilies and colloquial speech. The Protestant Episcopal service reminds its communicants that Matrimony signifies "the mystical union that is betwixt Christ and his Church," and commands "Those whom God hath joined together, let no man put asunder." In colloquial speech, we hear the wife referred to as the "better half," or less graciously, as the "ball and chain." These terms are not generally applied to one's mistress. She may be very difficult to be rid of, but she is never a "ball and chain." The husband conceives of himself as joined to the wife anatomically, at least in those cases in whom this fantasy plays a prominent role.

It is expressed, too, by the tendency to treasure one's wife, to fondle her and treat her tenderly, all the time considering her a very precious appendage, and certainly senseless! The manipulation of the penis is transferred directly to treatment of the wife in a masturbatory fashion. "Self-abuse" becomes, as it were, "wife abuse." Thus it is a vehicle for true sadistic expression, whether by teasing, or more violently, and having as its aim the production of an orgastic equivalent in the wife —tears, a temper tantrum or some other manifestation of loss of control. (While the husband maintains his!) More favorably, albeit less often, the aim is the production of a true orgasm in the wife, although for the husband this represents merely an ejaculation of his girl-phallus, not a true attempt to give pleasure to another person.

The fantasy acts as one of the determinants in the preference for certain variations in the sexual act, as, for example, that in which the woman sits on the man, or in which there is emphasis on fondling or other means of treating the wife like a phallus. This was particularly well brought out in one of Oberndorf's cases.[4] Here this sexual practice played a prominent role in a wife with pronounced body-phallus fantasies.

It is generally, perhaps universally, expressed by a type of dependence which is notably lacking in respect. The husband feels he cannot do without his wife, but he is hardly ready to dignify her by admitting that she has any more than material or even mechanical importance. Her presence as an object of reassurance, manipulation and exhibition are obvious; but she is not appreciated as a companion or helpmeet.

This fantasy is, of course, only one among many determinants, but a rather important one in this type of man. Many other unconscious sources, genital and progenital, may be found, all of them important in various degrees. This particular fantasy can, at any rate, be traced to its clinical manifestations with a high degree of specificity, and conveyed in interpretations.

Tracing it back to its origins in earlier phases of development is, however, a much more difficult task, and the results are less convincing. In the practice of psychoanalysis, increasing penetration of the most infantile elements of the personality has led to greater difficulty in establishing specific interpretations which can be validated, along with the tendency to construct general theories which can be applied all too broadly.

For this reason, among others, it remains something of a problem to discover why this particular fantasy developed and retained such power in these individuals. It seems likely that the fantasy is universal in men, as its converse may be in women, but we must assume the presence of some group of factors to account for its exaggerated importance in these patients.

These determinants are of complex origin, derived from the problems of identity, reality and sexual development in the young child, and may be traced to the baby's earliest relations with the outer world of feeding and mother. In all cases the fantasy appears to operate as a defense against fear of imminent loss, whether of the mother or as a part of one's own self. This is discussed at length in another paper on the subject.[7]

One of the most interesting aspects of this fantasy-complex is that, although it is basically very regressive and although it may be associated with very severe sado-masochistic behavior in marriage, it may also be manifested in ways which simulate adult love. A man who treasures and fondles his wife is not necessarily thought of as being a neurotic and he may not be. It is not the fantasy itself which makes the neurosis. It is the prominence of it, the ways in which it is carried over in behavior, if it is, which make for trouble. Nevertheless, in the "well adjusted" man one would expect that other concepts and attitudes about his wife would overshadow so regressive a one as this.

This fantasy ("my wife is a part of me, my phallus") is after all only one of a very large series of fantasies that husbands may have. There are more familiar ones—the cliché "my wife is my mother" or "my wife is my sister." In all probability, fantasies such as "my wife is my baby," "my wife is a pure and innocent princess whom I must rescue from evil monsters (parents)," "my wife is a castrated boy," "my wife has a hidden penis" are equally common. There are a host of others. The more common ones are to be discovered in all men, the least common perhaps only in those who are clinically ill, although I suspect that all husbands share most of these fantasies to some degree. They co-exist in the individual, they are encouraged or discouraged by life experiences and the character of the wife as well as the husband. They may exist often in odd and incongruous forms. For example, husbands rescue "helpless maidens" who often turn out, in fact, to be very competent, even domineering, viragos. Unconscious fantasies are not, after all, readily corrected by mere experience.

Such phenomena are not in themselves the cause of marital difficulties. They are rather the expression of the individual's attempt to satisfy important emotional, or better, instinctual needs. To what extent do they have more than theoretical importance? If we accept the thesis that marital problems are to a considerable extent, although certainly not completely, the reflection of underlying personality disturbances in one or both partners, we ought also to agree that a scientific, as contrasted with an intuitive, approach requires understanding of these disturbances in both their genetic and their dynamic aspects, that is, what causes them and how they operate in the personality. It is obviously not necessary, nor is it feasible, for everyone who wishes to help a distressed couple to be familiar with their specific unconscious fantasies. The results attained by sympathy and intuition are not to be scorned. But insofar as we recognize the need to search further into such problems and to achieve more effective results, we must look for mechanisms and causes.

A fantasy such as the one I have described may be considered an intermediate step in a psychological process rather than a fundamental cause. It is presented in this spirit: that it is only one of a large series of fantasies which act as the vehicles for the unconscious instinctual derivatives and defenses of the individual. Knowledge of such fantasies is vital as part of the therapeutic process of analysis of the individual, and is necessary in order to allow the analysand to attain an adequate degree of control over his infantile demands, to render his behavior relatively free of the need to satisfy such primitive wishes in literal and destructive forms.

References

1. Fenichel, Otto, "The Symbolic Equation: Girl-Phallus." *The Psychoanalytic Quarterly*, XVIII, 303-324, 1949.

2. Ferenczi, Sandor, "Gulliver Fantasies." *International Journal of Psycho-Analysis*, IX, 283-300, 1928.

3. Lewin, Bertram D., "The Body as Phallus." *The Psychoanalytic Quarterly*, II, 24-47, 1933.

4. Oberndorf, Clarence P., "Psychoanalysis of Married Couples." *Psychoanalytic Review*, XXV, 453, 1938.

5. Reich, Annie, "A Contribution to the Psychoanalysis of Extreme Submissiveness in Women." *The Psychoanalytic Quarterly*, IX, 470-80, 1940.

6. ————, "Narcissistic Object Choice in Women." *Journal of the American Psychoanalytic Association*, I, 22-44, 1953.

7. Stein, Martin, "The Marriage Bond." *The Psychoanalytic Quarterly*, April, 1956.

Analysis of Reciprocal Neurotic Patterns in Family Relationships

BELA MITTELMANN, M.D.*

NEUROTIC CIRCULAR INTERPERSONAL REACTIONS among individuals who are in close and intimate contact are an important component of the neuroses, and may follow one of several patterns.

Such patterns may exist between parents and children, between business partners or employer and employee, between siblings[2] and between sexual partners. Psychological factors were obviously at work in many of these relationships in the original choice of a "mate."

The complementary reactions described in this chapter were reconstructed from observations of twenty-eight couples, eight concurrently analyzed, seven in concurrent prolonged psychotherapy, seven couples with one mate in analysis and the other receiving briefer therapy, and six couples with one mate

---

* Associate Visiting Neuropsychiatrist, Bellevue Hospital, New York.

being treated and the other seen a few times or known only from the patient's reports.

The Dynamics of Neurotic Interrelations

Despite the misery and suffering inherent in some marital relationships, they do manage to fulfill certain vital needs for both participants. However, the emotional patterns of the mates complement each other in such a way as to perpetuate their pathological reactions through an intrapsychic vicious circle of reactions.

Complementary patterns between marriage partners, to be listed below, may be found in combination. The nature and intensity of disturbances in genital function, as well as the relation of childhood fixations to the dominant adult behavior, may vary with the identical complementary pattern.

One such pattern between marriage partners is characterized by one member's being aggressive, sadistic, out to humiliate and hurt his partner, thus relieving his own anxiety aroused by the relationship. The other member is chiefly dependent, submissive and enduring. The wife or husband may play either role, both of which are charged with conflicting trends, leading to outward disturbances.

Another common pattern combines an attempt at self-sufficiency through emotional detachment on the part of one partner (usually the man), with an intense, open demand for love on the part of the other (usually the woman). When the woman's violent demand for love and support arouses the man's fears, he becomes even more detached, while she evaluates this as a humiliating rejection. Both partners project the guilt arising from their mutually aggressive attitudes and blame each other for their difficulties.

Such a marriage often takes place because of a common illusion. The woman, in search of a strong mate on whom she can lean, evaluates the man's detached calm as strength. The man appraises the woman's vivacity, particularly if she has a vocation, as independence, and thinks she will make no demands on him either for support or for open affection. Early in the marriage the first difficulties are apt to arise through the growing disillusionment of both parties in the sphere of genital satisfaction. Later the marital difficulties are precipitated by external events.

Another complementary pattern may consist of a mutual attempt at domination, coupled with a violent defense. In such stormy relationships there may be as many as eight major quarrels per day. Both partners are critical of each other, feel constantly insulted and humiliated, and set out to humiliate each other in turn. They are both in need of affection. Their intense longing for dependency is either unconscious, resulting only in greater sensitiveness, or it is not recognized in its full intensity and is presented as a frustrated genuine desire for affection. Each wants to win a complete victory over the other at any cost, and because of strong dependency needs each is alarmed at the prospect of losing the other.

Neurotic illness with a plea of helplessness on the part of one mate and an attempt at extreme considerateness on the part of the other creates another type of complementary pattern. The difficulty here arises from a number of factors. The partner suffering from the frank neurosis expects the omnipotence and perfection of his mate to relieve his suffering. Always disappointed in these expectations, he expresses his unconscious resentment through depression and an exacerbation of symptoms.

The considerate mate, on the other hand, is not so motivated by love alone, but is forced to be extremely patient because of a lack of self-confidence. At the same time he is strengthened by the idea of helping a weaker individual. Ultimately, however, despite the self-imposition of endless restrictions on his activity out of consideration for his mate, he fails. The resultant loss of self-confidence, reinforced by an intuitive recognition of the sick person's criticism and disapproval inevitably leads to intense resentment.

A more complex complementary reaction pattern is seen in a syndrome where periods of helplessness and suffering are followed by periods of intense self-assertion on the part of one mate, and periods of shouldering responsibility followed by a disappointed desire for love and support on the part of the other.

In all of these complementary patterns both partners may obtain a measure of satisfaction and safety. Mutual identification between the partners can play a significant supportive role in two ways:

1. Both mates may find security, satisfaction and increased self-esteem through a mutually over-idealistic approach to life.

2. Helplessly dependent, each considers the other a haven of refuge, and by helping the other helps himself. The feeling of helplessness then may be renewed through identification with the helpless partner. Also, resentment is engendered when one partner does not live up to expectations and thus makes the other's deficiencies again evident. Furthermore, one partner assumes as his own those deficiencies manifested by the mate and thus feels guilty and inadequate.

Fears of frustration, condemnation, abandonment and attack arise from one partner's own impulses of hostility, or submissive dependency needs, as well as from the other partner's behavior. The reactions are likely to be most intense when intrapsychic distortion and reality coincide. Thus, patients consider a partner's hostile over-reaction as proof of the catastrophic results of self-assertiveness.

The habitual attitudes of a spouse may contain the original reaction to a parent or other infantile experience, and this important concept is dealt with in detail throughout this volume. Thus the demand for gratification of frustrated longings for dependence may have persisted throughout the patient's life. Without insight into its nature, extent, and unconscious motivations, it may dominate marital behavior. Current attitudes of dependence, hostility, sexual needs, anxiety, guilt and infantile experiences are interrelated in complex complementary reactions. Rejection by the mate leads to anger, then to anxiety and guilt. This revives infantile helplessness and dependence, with fear of genital activity and reinforcement of infantile striving, which, in turn, reinforces current anxiety and guilt and leads to anticipation of further rejection.

COMPLEMENTARY REACTIONS IN PARENT-CHILD
AND SIBLING RELATIONSHIPS

A brief survey of similar patterns between adults and children and among children is relevant because such patterns bear heavily on the relationship between parents.

The parents' behavior toward the children may be consistent or inconsistent and alternating. Both parents may have essentially similar attitudes toward the children or they may be in conflict with each other, and one mate may even use the child in the struggle with the other mate. The child's general behavior and personality formation may be in the direction in which the parent or parents push him or he may be in rebellion against it.

Systems of interrelation may develop on the basis of any of these variations. The relationship between siblings may be similar to those described for adults, except that the relationship with the parents also plays a significant role either in the form of opposing the mutual attitudes of the children or favoring them.

The simplest situation exists in what may be called the "anxious family." Here both parents are overanxious about themselves, each other and the children. The parents' anxiety may manifest itself mainly in the form of hypochondriacal concerns or oversolicitous, overprotective behavior. The children go along with this excessive anxiety and each member's anxiety is reinforced by that of the others. This also acts as a sign of reassuring love and care. Of course anxiety is not the sole emotional force in such interlocking relationships. But all other impulses, particularly aggression and at times sexual impulses, have a tendency to turn into anxiety.

On the other hand, the parents' anxiety may be met by rebellion in the child, mainly because it is experienced as a restricting force. The parents then become disapproving and punitive. This criticism and punitive attitude may aggravate the child's rebellion, which in turn increases the parents' anxiety and disapproval.

Similar conforming or rebellious patterns in the child may be observed in interlocking relationships with overambitious, perfectionistic or otherwise oversystematic parents. Such parents regard the child's conformity with their rules as proof of the correctness of their attitude. This, in turn, reinforces the child's behavior. However, if the child rebels, disapproval is added to the parents' driving insistence. Anticipating a variety of future disasters for the child, the parents reinforce their demands, again increasing the child's rebellion which may show up in open defiance, passive failure (as in schoolwork) or in somatic symptoms.

Other reactive circles may obtain between rejecting, critical parents and the child if the child rebels or fails in his functions. The parents then consider their rejecting attitude as justified, which in turn has a reinforcing effect on the child's symptomatology. If the child of overpermissive parents develops tantrums —partly as a result of lack of acquired control and partly as a result of guilt and anxiety over aggression—the parents may interpret the symptoms as a further reason for overpermissiveness, again establishing a circular pattern. Sometimes there is a combination of patterns, with excessive anxiety and overpermissiveness as well as excessive ambition for the child.

Inconsistent behavior on the part of the parent may facilitate inconsistent behavior on the part of the child, the two patterns again reinforcing each other. A parent may be very permissive, loving, close and then overambitious, driving, scolding, critical and punitive. The child may meet one of these attitudes, namely the excessive driving, with rebellion and partial failure, which would increase the parent's driving behavior. Beyond a certain point, however, the child may turn on his charm and affection, which would then result in the parent's suspension of all punitive measures and lead to excessive permissiveness. This may temporarily heal the breach, but the parent's demands would not be met on later occasions, leading to a recurrence of the cycle.

Similarity of orientation toward the child may be an important bond even between neurotic parents. While vicissitudes in handling the child may also act as a bond, more often they intensify disagreements which occur for other reasons of a neurotic nature.

In regard to differing attitudes on the part of the parents, if one parent is overanxious and the other not, the child's anxiety may be considerably lessened, provided the relationship between the parents is essentially a good one and the disagreement does not flare into open quarrels. Even so, the child may then have conflicting identifications and may be critical of either parent. If the quarrel between the parents is open, the child's anxiety increases, the conflicting identifications are more intense and to this is added fear of abandonment and confusion of self-identity. The child's anxiety may serve to reinforce the anxious parent's anxiety and the combative disagreement of the other parent.

The sexual element may play a significant role in all of these patterns, or it may be a dominant pattern in itself. There are families in which every member behaves seductively toward every other member, reinforcing such behavior on the part of each. In some instances, one or both of the parents may be seductive and the child may go along with it. Or the child may be seductive and the parent accept it or react to it with anxiety and conflict. The latter behavior may either discourage the child or increase his seductive behavior.

Seductiveness may be used by either parent or child to satisfy his need for support and love, or to placate the critical or rebellious member. This then leads to alternating and inconsistent cycles of the type previously discussed, except that here the sexual-seductive behavior takes the place of or is combined with overpermissive behavior.

It may be mentioned that in almost all these relationships there is a positive element both for the parent and the child. For the child it is one of gratification of dependent needs. For the parent it is self-centered (narcissistic) gratification. Unless they are excessive and gross, these aspects may be lost sight of in the total pathology.

Among siblings, the most common interlocking pattern is rivalry. However, either as an outgrowth of rivalry or as a result of problems with the parents or absence of parents, other significant interlocking patterns are to be found. One of them is that of dependency, where one sibling is considered strong by the other, who is obedient, submissive and dependent. There may be a sexually seductive relationship between them, both of them participating equally or one acting seductively toward the other. The bullying older sibling may be the seducer, with the younger one submitting in order to placate him.

The parents may play a role either as sought-for sources of exclusive affection in cases of sibling rivalry or as targets of aggression if the other sibling is favored. The siblings may band together, partly in the form of mutual dependence and seduction, in a common front against the parents. Of course the pattern between the siblings is always related to and in turn affects that between the parents.

Therapy

The therapeutic management of individuals involved in complementary neurotic relationships is handled in actual practice in one of the following ways:

1. Only one of the mates undergoes treatment.
2. The therapists are separate for each member and do not communicate with each other.
3. The therapists are separate for each member but are in communication with each other.
4. The same therapist simultaneously handles all members of the interlocking relationship.

In about fifty per cent of the cases, it would appear that any of the four procedures may be used. In about twenty per cent, separate therapists communicating with each other obtain enough information to take care of the particularly troublesome interlocking problems. In another twenty per cent, the management by the same therapist would appear indispensable for full therapeutic effectiveness. In about ten per cent, treatment by separate therapists is indicated.

Treatment by therapists not in communication with each other, or where only one of the mates is being treated: If the complementary reactions are not of a serious nature, it will suffice to explain to the patient on the basis of daily reactions how his behavior is affecting the mate and how he in turn reacts to the behavior of the partner. Such interpretations are needed because the patient's neurosis has one of its strongest points of anchorage in the family relationship.

If the patient's analysis is successful, the complementary reactions of the marriage partner or of the child often disappear without direct therapeutic measures. The patient gradually corrects the biased presentations of reality incidents. Also, he fills in gaps in his report of daily events, recognizing for example that some remarks of the mate are the result of anxiety aroused by the patient's own complaints and not of the mate's tendency to criticize and dominate.

The important question arises as to the point at which the complementary reactions can best be taken up during the course of the analysis. If the patient is putting the blame for much of his difficulty on the mate early in the analysis, some discussion about the mate's behavior and the patient's reaction should be taken up early. At this period, however, it is impossible to discuss the details and dynamics of the complementary reaction. Before this can be done, the patient's transference reactions and unconscious childhood wishes must be analyzed to a considerable extent.

Regardless of when the complementary reactions are discussed, the patient may react with relief because of insight and implied reassurance. On the other hand, he may react to them as criticism and an attempt on the analyst's part to take the mate's side. These reactions can be resolved in terms of the patient's habitual characterological traits, genetic determinants and transference reactions.

It may be mentioned that the overwhelmingly prevalent custom at present in the therapy of adults is that the therapists treat only one of the mates without communicating with each other. Apart from the nature of the respective neuroses of the partners involved, the treatment of only one partner is apt to be satisfactory if he is the emotionally dominant one. The following case will illustrate this concept.

A thirty-three year-old male psychologist entered psychoanalysis for training purposes, presenting himself as essentially symptom-free. Rather early in treatment, however, it appeared that he exhibited symptoms of a compulsion neurosis in that he prayed compulsively in stress situation, turned his shoes toward the door when retiring, and did not initiate certain actions on certain days to avoid bad luck.

He also suffered from temper outbursts, most often directed toward his wife, in which he was critical of her either for not preparing their meals properly, for handling their two children poorly or for being forgetful. At first he considered these bouts of criticism perfectly justified. Later he realized they were much in excess of the alleged provocations. It further turned out that he was constantly asserting his superiority and infallibility, wanting humble self-depreciation on her part as the price for his forgiveness and affection.

The wife conformed to this role. She considered herself inferior to her huband, constantly leaned on him for advice and made no decisions on her own. This made him more critical. Later in his analysis he realized that many of the qualities he was so critical of in his wife were the result of his own doing. Her lack of decisiveness and initiative was in part due to the fact that he considered only his own judgment reliable. Her forgetfulness was often due to his having been critical of her previously, thereby putting her under tension.

At one point in the treatment, after transient improvement, the patient's critical attitude toward his wife grew more intense. As a result, the wife started to rebel, whereupon the patient talked of divorce. Then the wife developed new anxiety equivalents, such as the involuntary shaking of one hand. When the analyst pointed out emphatically that he was contributing to his wife's difficulties, the patient's temper subsided and he began to apologize to his wife after outbursts.

It became clear that he had a similar need to be master of situations in other set-ups as well, his relationship with his mother having a great many parallel features. His mother had extolled him and still did so over any other person, including the patient's father. The patient revealed that he used to engage in similar outbursts toward his mother and father whenever they opposed or criticized him in the slightest, or when they did not do exactly as he wished. In the course of three years of analysis this patient's symptoms cleared up without the therapist's having had direct contact with the wife.

This case also illustrates the fact that a marriage containing strong neurotic complementary factors (in this instance the husband's need for domination and superiority and the wife's need to lean and submit) can be rather successful. One element in this success was the dovetailing of the couple's genital drives.

The fact that the wife desired long clitoral stimulation, after which she could reach vaginal orgasm in about a minute, coincided with the husband's somewhat premature ejaculation.

The two mates have separate therapists who are in communication with each other: Except for a few publications,[3] there is very little in the literature about such procedure and no systematic studies are available. In the author's experience, communication between therapists may either be very occasional or may take place at regular intervals, perhaps once or several times a week. The following case is an example of this procedure:

The husband had entered analysis with the author for an anxiety state a number of years earlier. Initially, he also suffered from a temporary potency disturbance. It was characteristic of his history that he would fall deeply in love with a woman, consider her quite perfect and marry her. Difficulties would arise in the marriage, whereupon he would fall in love with another woman, also regard her as being perfect, divorce his wife and marry the second woman. His current marriage was following a similar pattern, but due partly to external circumstances and partly to his analysis, he did not go so far as divorce. Instead, his wife also entered treatment.

The wife was frigid and found it difficult to reach orgasm even through clitoral stimulation. This was very disappointing to the husband who regarded it as a failure on his part and a rejection by the wife. She also suffered from a moderately rigid personality and a tendency to anxious somatic complaints, such as palpitations, fatigue, etc., for which no gross organic disturbance could be found. For about a year she had resisted the idea of going for therapy, feeling she ought to be able to take care of her problems herself. When she finally consented, she said she wanted a therapist other than her husband's analyst. She would be embarrassed with the husband's analyst.

After the wife had been in treatment for about a month, her analyst and the author had a discussion. The wife's analyst asked whether there was any truth in the wife's vague stories about her husband's infidelity, or whether she was really suffering from a paranoia of jealousy. Actually, the wife was trying very hard to deny patent evidence of the husband's infidelity. In this case the wife's therapist welcomed early clarification of the fact that his patient was exhibiting hysterical denial, rather than signs of a borderline psychosis.

Obviously, the therapist in such cross communication must remember very clearly what information he has received and from what source.

The same therapist simultaneously handles two or more members of the same family: The technical aspects and the indispensability of this procedure for adequate therapeutic effect in some situations has been discussed in previous articles.[4, 5, 6] In some cases of difficult complementary neuroses, one therapist for both mates is advantageous for several reasons. The therapist obtains information in the course of the treatment which makes him more perceptive to the problems of each mate. Also, he can see the bias with which one or both of the mates interprets significant facts and can discover the relevant omissions. In periods of severe stress for one of the mates, he can if necessary modify the type of behavior on the part of the other which was intensifying the tension of the first.

Simultaneous treatment of husband and wife by the same therapist seems indispensable when one or both of them distort the situation in all honesty and do not correct this distortion in the course of treatment even after several years. In such instances, even after the analyst has knowledge of certain facts, he cannot elicit the information from the mate until he actually confronts him with the statement obtained. Of course, this has to be done with the mutual consent of both mates. Such a case was described in a previous article.[6] The following is a brief sketch of an essentially similar problem: A couple was referred to the author for treatment. They quarreled with each other several times a week, and there were periods of strong tension between them. Each gave a fairly convincing story justifying his own complaints.

The wife suffered from asthma. Early in her analysis she repeatedly stated that her attacks had no connection whatever with the emotionally charged situations. Six months later, when the analyst questioned the husband about this matter, he said it was clearly understood between his wife and himself that her asthma got worse periodically whenever she became overly excited. Confronted with this statement, the wife at first denied she had ever said there was no connection between her asthma and emotionally charged situations. She had merely said that emotional problems were not the exclusive cause of her asthma. This of course was a retrograde distortion which she herself admitted after a brief discussion.

For a while after this she reported major fluctuations in her asthma in connection with emotionally charged situations. Then she stopped reporting on them. After the husband mentioned some situations to the analyst, the wife again went through her former distorted excuses. Finally, about seven months after first being confronted with her husband's views on the subject, her asthma improved considerably.

Of course, interpretations on the dynamics of the asthmatic attack were possible only when the patient was willing to consider their psychosomatic character. Further, the wife's asthmatic attacks were a sensitive gauge of her neurotic involvement with her husband and children. These neurotic trends were partly inaccessible when she shut her asthma out of the therapeutic process.

It should be noted that while some of the most striking material withheld by such patients is not deeply unconscious, they are actively reluctant to recognize its significance. One woman judged the effectiveness of her analysis by the yardstick of attaining full vaginal responsiveness. Yet in the course of four years of treatment she had not breathed a word about the fact that she herself never prepared the vaginal diaphragm but always expected her husband to do it. Without such material emerging in the analysis, the wife's orgastic responsiveness could not be fully restored. In this particular case the necessary information was obtained from the husband's analysis.

Such withholding of facts may be combined with exaggeration and distortion of trivial acts of the mate when there is constant quarreling between the couple. Occasionally, this distortion may go so far that the therapist does not get a clear picture until he holds a joint interview with both mates. Otherwise, of course, the mates are always seen separately.

Generally, in the treatment of children as well as psychotic adults the therapist is in constant contact with some key informant in the environment, although, as a rule, the treatment of this key informant, if needed, is conducted by another therapist. Some investigators, however, have found that the treatment of the child and the adult by the same therapist is of distinct advantage, even without special complications in the situation,[1] and even indispensable where the child is suffering from a psychosomatic disorder.[7]

There is another advantage in the use of cross information where each mate reacts intensely to some behavior on the part of the other. One of the pair may be over anxious and try to coerce the mate into a certain course of action. The mate may take this as criticism and an attempt to dominate. Such reactions may be so frequent and intense that they exclude almost any other topic from the therapeutic sessions. The therapist may then find it effective to inform the "injured" party about the true nature of the mate's motivation. In some cases this kind of correction and implied reassurance is of considerable benefit. Cross information is therapeutically useful when it supplements the patient's coping ability (ego strength).

In cases where the concurrent handling of both mates seems indispensable, the accent is always on the current situation. The treatment cannot proceed satisfactorily without clarifying the current situation in each hour and then going on to other material. In all probability this would be the case even if these patients were in analysis by themselves. However, concurrent treatment does contribute to the emphasis on the current situation.

TRANSFERENCE REACTIONS

The analysis of neurotic difficulties involves patients' attitudes toward the therapist, and similarities and differences in the mates' behavior toward the analyst are instructive. One spouse may be taciturn and self-effacing, the other friendly and talkative. They may both speak quietly, rarely associate freely and lapse into silences. Both partners may show boundless faith in or superiority toward the analyst. Or each of them may manifest contrasting attitudes. Such observations enable the analyst to understand the mates' reactions toward each other and to judge the reality of their mutual problems.

The following attitudes may arise in either one or both married partners who are being treated by the same analyst:

1. Concern as to whether the analyst agrees with one patient's over-evaluation of the mate or whether he considers him or her deficient.

2. The fear that the analyst is siding with the other partner.

3. The wish-fulfillment fantasy that the analyst will change the mate by magic, and thus save the patient the need for working out his own conflicts.

Concurrent treatment of marital problems may also result in one partner blaming the analyst for the assertiveness of the previously submissive mate, with perhaps the complaint that the analyst deprived the complaining partner of any defense against such attacks. Fear of divorce may arise in connection with the liberation of repressed aggressive or submissive impulses. (In practice such fears are unfounded. Concurrent analysis seems to save some marriages that would otherwise have gone on the rocks.) There may be complaints about the destruction of over-idealized values attached to the marriage.

A positive note is struck in the concurrent analysis of married couples when one of the patients feels that by changing the mate the analyst has altered a reality which formerly presented insurmountable difficulties. The correction of these difficulties relieves the patient's helplessness, increases his confidence in the therapist, removes some of his most important rationalizations and enables him to face his other problems.

THE THERAPIST'S ORIENTATION

Not all therapists are at home in treating both mates concurrently. A reliable memory is an essential requirement; the therapist must know from which mate he got what information. The therapist must be able to withstand the rivalry and attack, at times sustained, of both mates. The therapist must be impartial, but not neutral. He must be able to take a clear-cut stand in critical situations as to who contributed what to a dispute, otherwise quarrels get out of hand. In other words, the therapist should behave like a benevolent, impartial and firm parent.

CONTRAINDICATIONS FOR CONCURRENT TREATMENT OF MARITAL PARTNERS BY THE SAME THERAPIST

Serious doubt on the part of the therapist as to whether the marriage may be saved is not necessarily a contraindication for concurrent treatment. If in the course of treatment one or both partners decides to dissolve the marriage, and continuing with the same therapist is too painful a reminder of this fact, the patient can always transfer to another therapist. However, if one of the partners is paranoid in the sense of being ready to blame everyone else for his problems, and would therefore probably blame the therapist for the termination of the marriage, it would seem wiser to have different therapists initiate treatment.

Frank paranoia in itself, however, may not be a contraindication to simultaneous treatment. In fact, in some instances, the social adjustment of the patient can be accomplished more readily by the same therapist who treats the mate.

Concurrent treatment by the same therapist also seems contraindicated if one of the mates objects to going to the same therapist as the other. If such a patient were advised to go to the same therapist, he would refuse to be treated at all and would hide all other resistances against treatment behind the cloak of his objections to concurrent treatment.

Despite these considerations, the author can see no objection to the therapist of one mate seeing the other mate in a few interviews for evaluation and clarification. Nor can there be any objection to the therapist's starting the other mate's treatment on a trial basis and, should contraindications develop, referring him to another therapist. In the author's experience it does not matter whether both mates start treatment at about the same time or one well ahead of the other. Nor does it matter if one of the mates starts treatment with another therapist, and later transfers to the therapist of the other mate.

Summary

Complementary patterns are an important aspect of the neuroses of patients in intimate contact with other individuals, particularly in marriage. Briefly, the following complementary patterns are seen in adult relationships:

 1. One of the partners is dominant and aggressive; the other, submissive, passive and masochistic.

 2. One of the partners is emotionally detached; the other craves affection.

 3. There is a continuous rivalry between the partners for aggressive dominance.

 4. One of the partners is helpless, craving dependency from an omnipotent mate; the mate is endlessly supportive.

 5. One of the mates alternates between periods of dependency and of self-assertion; the other between periods of helpfulness and of unsatisfied need for affection.

The pattern of the infantile intrafamily constellation in certain instances may be along the same lines as the adult pattern. In other cases several shifts may have occurred, and the adult pattern may even be the opposite of (although of course dynamically related to) the pattern that prevailed in childhood.

Complementary patterns may play a major role in the therapeutic problem. Some of the complementary reactions, including the genital ones, afford relief to the patient; others are such as to perpetuate and renew pathological reactions. It is a useful and sometimes indispensable therapeutic measure to concentrate the treatment on these complementary patterns.

In many patients this can be done adequately even if the patient is treated by a therapist who is not in contact with the mate. In other patients the complementary problems can be adequately handled if the separate therapists of each mate are in frequent communication with each other. In still other patients it seems indispensable that the same therapist handle both mates.

Transference reactions include concern as to whether the therapist favors the mate, whether he values the mate as highly or as poorly as the patient, fear of divorce or separation and defense of unattainable marital ideals. In concurrent treatment of both mates by the same therapist, the current situation must be clarified continuously and, in case of intense and sustained quarrels, the therapist must be impartial but not neutral. This means he must make clear to each party the extent to which his reactions represent problems and their inevitable effect on the partner.

Concurrent treatment seems inadvisable if either mate is reluctant to go to the same therapist as the other primarily because of a feeling of shame and embarrassment; if either mate demands information about the other from the therapist; and/or if the marriage seems doomed to failure, for which one of the mates would be likely to blame the therapist.

References

1. Burlingham, D. T., "Present Trends in Handling the Mother-Child Relationship During the Therapeutic Process." *The Psychoanalytic Study of the Child*, Vol. VI, 31-37. New York, International Universities Press, 1951.

2. Lehrman, P., "An Initial Visit." *Journal of Nervous and Mental Diseases*, LXXIV, No. 1, 1931.

3. Martin, P. A., and Bird, H. W., "An Approach to the Psychotherapy of Marriage Partners: The Stereoscopic Technique." *Psychiatry*, XVI, 123-127, 1953.

4. Mittelmann, B., "Complementary Neurotic Reactions in Intimate Relationships." *The Psychoanalytic Quarterly*, XIII, 479-491, 1944.

5. ————, "The Concurrent Analysis of Married Couples." *The Psychoanalytic Quarterly*, XVII, 182-197, 1948.

6. ————, "Simultaneous Treatment of Both Parents and Their Child." *Specialized Techniques in Psychotherapy*, G. Bychowski and J. L. Despert, eds. New York, Basic Books, Inc. 1952.

7. Sperling, M., "The Role of the Mother in Psychosomatic Disorders in Children." *Psychosomatic Medicine*, XI, 377-386, 1949.

Sexual Problems in Marriage

VICTOR W. EISENSTEIN, M.D.*

CLINICAL EXPERIENCE indicates that a good sex life does not assure a happy marriage, nor do sexual difficulties necessarily cause marital breakdowns. Sexual symptoms occur in the context of either manifestly harmonious or frankly discordant relationships. There is no question, however, that happy marriages, by and large, are marked by a greater degree of sexual satisfaction, while unhappy marriages have a much higher incidence of sexual conflicts.

The sexual symptom is only one of many indicators of individual emotional problems which find numerous avenues of expression in a marriage. Disturbances in the sex life are related to the individual neurosis, to the neurotic choice of a mate, and to the resultant type of neurotic interaction between partners. (Sexual difficulties due to advanced age, organic changes in the sexual organs, or other somatic conditions are not under consideration in this study.)

Psychosexual immaturity is necessarily a feature of the un-

---

* Attending Psychiatrist, Chief in Psychiatry Clinic, Lenox Hill Hospital, New York City.

conscious conflicts of neurotic individuals. Emotionally immature people are incapable of experiencing satisfactory interpersonal intimacy, including heterosexual activity. Their frustrations may be associated with other neurotic symptoms, such as
depressions, phobias, compulsions, hypochondriacal anxieties,
or hysterical conversions, expressed as headache, cardiac palpitation, diarrhea, or other bodily manifestations.

Hypothetically, the so-called mature or normal person is
endowed with a full capacity for love. Such an individual has
resolved all his infantile conflicts in infancy, bringing to adulthood and marriage an integrated blend of tender and sexual instincts which enable him to undertake a lasting and satisfying
relationship with a mate. The degree to which this ideal conception of maturity falls short of being attained is seen in the
wide incidence of neuroses, and in the vast numbers of unstable and discordant marriages, only a fraction of which end in
separation or divorce.

Since the marital union is a dynamic rather than a static relationship, circumstances are bound to arise through which unconscious infantile conflicts may be revived and find expression
in sexual disorders. Premature ejaculation or impotence in the
male, for instance, may appear after the wife has given birth, or
has undergone surgery. Or it may occur following a death in the
family, or after an injury to self-esteem in a business or professional matter. What is popularly known as sexual incompatibility may arise when the partner acquires a new meaning in the
unconscious life of the affected individual, thus reactivating
deep-seated conflicts centered around a parent or sibling in early
childhood.

With tensions and hostilities mounting in a discordant marriage, refusal to cohabit and lack of sexual interest in the partner are symptomatic of the over-all disturbance in intimate living. Yet even when other aspects of the marital adjustment are
relatively good, sexual disturbances may arise or become accentuated.

Frequently, partners who are devoted to each other and live in harmony, but suffer from sexual difficulties, will seek the help of general practitioners, urologists or gynecologists for these inexplicable symptoms. Few such individuals recognize that their sexual problem is a sign of emotional disturbance calling for psychiatric help.

While affording the normal adult a pleasant feeling of satisfaction and relaxation, coitus can be a source of frustration and tension for the neurotic individual. Because of unconscious conflicts, some archaic wishes in the neurotic person are unsatisfied and indeed cannot be satisfied through sexual intercourse.

Many people enter into marriage with such intense feelings of guilt about sexual matters that they cannot tolerate intercourse. They react to coitus with insomnia, depression or irritation which expresses itself subtly a day or so later in regard to non-sexual matters. As Freud first pointed out, many men are impotent with the woman they love and sexually potent only with "degraded" women.[5] Such phenomena occur in people fixated to incestuous fantasies, for whom sex and taboo have become inextricably interwoven.

The degree of fixation to infantile sexuality determines the type of sexual activity that is possible. If the unconscious meaning of the act causes anxiety or guilt, there will be interference with sexual capacity or pleasure. If the unconscious fantasies do not arouse conflicts, if they are relatively free of anxiety and guilt, no sexual symptoms will ensue. These factors can be readily observed in a study of the dreams which follow sexual intercourse.[2]

For purposes of convenience in examining the subject, the commonest sexual symptoms encountered in disturbed marital relationships may be listed as follows:

1. Frigidity
2. Potency Disturbances
3. Preference for Masturbation
4. Preference for Sexual Perversion

5. Hypersexuality
6. Lack of Sexual Interest

It should be noted that these categories are not mutually exclusive, but are in fact interrelated.

Frigidity

Frigidity denotes the inability of a woman to attain a sexual climax from coitus. It does not mean merely the lack of sexual approachability. The inability to attain orgasm from a heterosexual relationship may be partial or total and more often than not exists in relation to any marital partner.

Freud states, "The sexual frigidity of women . . . is still a phenomenon which is insufficiently understood. Sometimes it is psychogenic, and, if so, it is accessible to influence; but in other cases one is led to assume that it is constitutionally conditioned or even partly caused by an anatomical factor." [6] Bergler delineates eight varieties of frigidity due to unconscious conflicts.[1] Because of cultural factors which foster greater repression of female sexuality, frigidity in women is more commonplace than potency disturbances in men. According to Kinsey, "Between 36 and 44 per cent of the females in the sample have responded in a part, but not in all of their coitus in marriage." [8]

A diagnosis of frigidity is not always an easy matter, particularly from the patient's earliest description. Sometimes a patient is in treatment for months before there is any indication that something is wrong with the sexual adjustment. There are many variables which affect the diagnosis. For example, a woman may reach orgasm but only through perverse fantasies, such as fantasies of being beaten. Without these, she is psychologically frigid and is bound to express her discontent in other ways.

The clinician encounters a large gradient of incomplete orgastic reactions. A woman may on one occasion suffer from near total frigidity with vaginal anesthesia, and on another have an almost complete orgasm. Some women experience relatively strong excitement in anticipation of coitus, but lose their desire during the act. There are others whose lack of vaginal pleas- ure is manifest even in the absence of lubricating glandular se- cretion in the vagina during sexual stimulation. In the most in- tense degrees of frigidity, fear produces dyspareunia, or painful intercourse, in which the vaginal sphincter muscle closes so tightly that it is difficult if not impossible for the penis to enter the vagina proper.

In all types of frigidity, strong sexual excitement may be felt, particularly in the clitoris, but pleasurable vaginal sensa- tion ends before the involuntary contractions of the vagina, characteristic of true orgasm, can be achieved. The many kinds of perverse sexual fantasies or activities which are evoked in an attempt to attain orgasm will be discussed later.

The basic significance of many cases of psychogenic frigidity is the unconscious comparison of the husband with the father, and the symptom is but one of a chain of defenses against "dan- gerous" infantile sexual aims. Frigidity may be a distorted ex- pression of unconscious anxiety connected with repressed sadis- tic impulses, such as revenge upon the penis and the man, or with masochistic aims of being genitally injured.

Fear of losing control during sexual orgasm is not uncom- monly the original infantile fear of losing control of urine or feces, unconsciously transposed into the present. In such cases there is often a remarkable recovery from this aspect of neurosis when the woman discovers and works through the meaning of her symptom and is able to attain orgasm for the first time in many years of married life.

In many instances an unconscious rejection of the female genital is in large measure responsible for the frigidity. This estrangement from one's own body image is a sexual form of a conversion hysteria. In other instances, sexual pleasure is blocked because unsuccessful attempts are made to use the genital apparatus for discharging warded off pregenital impulses, notably oral and anal strivings.

The following case illustrates some of the deep meanings of frigidity:

A twenty-eight year-old woman, totally frigid, was married to a man who suffered from premature ejaculations. The woman's early childhood was most traumatic. Her father had been physically brutal to her, and following the birth of his son had rejected her almost completely. The mother, an anxious woman, had considerable difficulty in breast-feeding her, and scolded her for manipulating her genital after the birth of a baby brother. The little girl was dressed in white and not allowed to play like other children lest she soil her clothes.

At puberty, the patient had erotic fantasies in the street, the content of which was intensely masochistic—she had visions of herself being run over by a heavy truck and mashed to pieces. As an adolescent, she had difficulty relating to boys, and feared she might be homosexual. Up to the time of her marriage, the patient shared a bed with one of her brothers.

In her marriage, she avoided sexual relations with her husband by sleeping in a separate room. She considered sex "dirty" and was fearful of sexual excitement. Extremely ambivalent toward her mother, she unconsciously identified with her father. Her husband was unconsciously identified with a brother whom she envied and hated, but to whom she was also incestuously fixated.

Consciously, she repeatedly expressed contempt for her vagina. During sexual foreplay, the sensations became intolerable to her because she had an excrutiating sense of being tickled. She often fought to free herself in the midst of a sexual act. The patient identified her body with a phallus and experienced intense "castration" anxieties in various forms, chiefly hypochondriacal.

Her dream and fantasy life were filled with terrifying images of being bitten, pierced, slit or torn. She unconsciously equated her vagina with the mouth, and the penis with the breast. Tearing out the bowels of her father, ripping off penises, and pulling off mothers' breasts were variations on the constant theme in her emotional life. In her fear of orgasm, she retained an infantile fear of "bursting."

Terrified of touching her own genital, she had as little contact with it as possible, except to wipe it after urinating. She constantly felt dammed up. Sexual approaches by her husband only made her frantic, and on several occasions she struck him during intercourse. Despite the sexual conflict, however, the marriage endured for many years.

Sexual frigidity very commonly results in marital infidelity —the frigid wife searches for sexual satisfaction, the husband reacts to his frigid and unresponsive wife. Legally, infidelity is grounds for divorce. Psychiatrically, however, it is generally the acting out of neurotic conflict. The woman who fails to attain orgasm with her husband often tends to project the blame entirely upon him, and is either tempted to seek out another partner, or actually does so. Certain character types, trying to negate their frigidity, will compulsively indulge in sexual activity in a promiscuous way. An extramarital affair may yield a measure of satisfaction so long as there is the quality of forbiddance—the earmark of infantile sexuality.

Generally speaking, frigidity cannot be cured simply by changing partners or environment. To be sure, there are some cases of facultative frigidity where the woman is frigid with one partner but not with another. Here the difficulty stems from the fact that she is unresponsive with the mate who has assumed an incestuous significance in her unconscious life. A neurotic choice of mates is involved in such cases.

The motive behind adultery is frequently associated with a sense of compulsion. The accompanying neurosis may be relatively mild or quite severe, and only an individual evaluation will determine the recommendations for psychiatric treatment. Many marriages are salvaged by analysis of the underlying neurosis.

Nymphomaniac types are in a perpetual state of sexual excitement, and enter into relationships more or less indiscriminately. These are invariably frigid women, primarily of strong oral disposition—the "vampires." Marital infidelity on the part of the woman may also be the acting out of a compulsive need to abandon the partner as belated revenge upon the mother or father for certain disappointments or rejections, real or fantasied, in early childhood. Not uncommonly, frigidity is the expression of an unresolved masculinity complex.

Marital infidelity on the part of either partner does not necessarily denote psychopathology. However, the more primitive the personality organization, and the more infantile and impulsive the character formation, the more exaggerated will be the degree of "acting out" as sexual promiscuity.

Potency Disturbances

The average duration of intercourse is from one to five minutes from the act of insertion until the completion of orgasm. In most instances, ejaculation is achieved after some thirty to fifty frictional movements, lasting about three minutes. There are, however, wide individual variations. Some men can perform active coitus from ten to twenty minutes before ejaculating, particularly with moments of rest during the act. Certainly, ejaculation which takes place in less than one minute may be considered premature.

In complete impotence, there is inability to attain or maintain an erection during an attempt at coitus. A more common potency disorder is that of premature ejaculation, where erection is attained but ejaculation occurs either prior to penetrating the vagina or within a few seconds after insertion. The opposite condition is retarded ejaculation, where erection is maintained within the vagina for a half hour or more with inability to achieve ejaculation.

As in frigidity, premature ejaculation in the male may be only a transient manifestation in an otherwise satisfactory marriage. Or it may exist under certain specific conditions without denoting pathology in orgastic potency. A man who has a normal duration of intercourse, but only by evoking perverse fantasies, on the other hand, is actually suffering from a potency disturbance.

Neurotic potency disturbances are generally built on complicated developmental foundations. While the dynamics of unresolved oedipal conflict vary in different types, fixation to infantile attitudes operates in all such instances—whether the compromise is expressed in complete impotence, premature ejaculation or retarded ejaculation. The most stubborn types of premature ejaculation are frequently associated with marked homosexual disposition.

Infantile castration anxieties, exaggerated narcissistic tendencies, hostile, grudging or non-giving attitudes toward women, unconscious equation of ejaculation with soiling or urinating, and similar infantile attitudes commonly lie at the root of these disorders. The exaggerated defense against aggressive and sadistic fantasies having to do with stabbing or piercing the partner is frequently responsible for the failure to maintain an erection on penetrating the vagina.

Adequate treatment, even in younger men, may require several years of psychoanalysis. Transient potency disturbances, such as those following traumatic experiences, respond more readily. Such transient disorders may occasionally be observed during the course of analysis of character difficulties when the patient is reliving his original infantile experiences and anxieties.

The following examples may serve to clarify the meaning of a potency disturbance in terms of personal development:

A passive, feminine man experienced lack of sexual desire for his wife and suffered from premature ejaculation in intercourse. During a period when he was most incapacitated by severe work inhibitions, his sexual life consisted of rubbing his penis between his wife's buttocks. This behavior was also reflected in the way he avoided the main highways when driving his car, choosing the "back roads" instead.

This dream, which he had after intercourse, is revealing: "My wife and I were riding in an open roadster. My wife said, 'Not so fast!' I said, 'Don't worry, I can see behind me clearly!' All of a sudden I heard a siren. I noticed the policeman on a motorcycle carting an outboard motor. He drove up and I became aggressive. I said, 'Before you start giving me a ticket . . .' and here I gave him a word beginning with fair (fairy). My wife said, 'I told you so.' (Here the patient added a previously forgotten fragment.) I slipped him a bribe from behind." [3]

The patient's aggressive attitude was really a denial of his castration anxiety, evoked as a punishment for sexual activity. Slipping the policeman a bribe from behind is a self-castration to ward off punishment. This element is a condensed reversal of his fear of and wish for the penis from behind.

This man unconsciously sought to be mothered by his wife. Whenever she left home to pay a brief visit to her mother, he would fly into a rage. Frequently there were scenes in restaurants when he would get up angrily in the midst of a meal and go home, with his wife following. He felt burdened when his wife did not go out and earn a living, yet ashamed when she did.

In periods when he could function at work, he sought out sexual relationships with prostitutes who performed fellatio upon him. In this way, by reversing the roles, he unconsciously played the role of the feeding mother, while the prostitute, identified with his infantile self, was the child at the breast. The core of his neurosis centered around his mother's pregnancies when he was two and four years old, at which times he had envied the suckling infants.

Another case of impotence, this one more stubborn, was that of a young chemist whose childhood was marked by a drastic series of disappointments in his mother. This reached a climax in events having to do with an operation. When he was five years old, his mother gave him a package, told him she was taking him to a birthday party, and, instead, brought him to a hospital for an operation. Having been totally unprepared, he proceeded to fight off the nurses and the anesthetic. Following a tonsillectomy and a circumcision performed at the same time, he awoke bleeding from the mouth and with a painful penis.

After this trauma, he split the mother image into two parts—consciously remaining tender and devoted to her, while unconsciously seething with rage. As he approached puberty, he was disturbed by occasional homosexual fantasies. Later, he became aroused by erotic fantasies of savage women fighting with each other, and used pictures or cartoons of fighting women as masturbation stimuli. On one occasion in his adult life, he went to a prostitute. After an unsuccessful attempt at intercourse, he dreamed that "I was trying to get my mother on the telephone. Men kept cutting in on the line."

He married a severely repressed and rather prudish young woman to whom he remained devoted and considerate. In attempting intercourse, however, he invariably called upon his masturbation fantasies of the fighting women in order to attain an erection. Neither partner was seriously troubled by the residual potency difficulty when it came to maintaining the equilibrium of this marriage.

An impotent man, married to a frigid woman, told his analyst that at the time of his marriage he was able to attain firm erections and reach the point of ejaculation after about a dozen frictional movements in the vagina. However, in a few months, he began to have difficulty maintaining erections after three or four strokes. He blamed this on his wife's unwillingness to participate in intercourse at regular intervals. Nevertheless, he treated her like a frightened child who had to be indulged by having few sexual demands made on her. His other neurotic symptoms were fear of heights, fear of driving over bridges and gastric disorders of functional origin.

The patient's father and mother were both executives in an advertising concern. He is the older of their two children, his sister being two years his junior. Throughout his childhood both parents worked, leaving the children to be reared by a maid. He described his mother as an unkind, inconsiderate woman whom he feared rather than loved. His other early attitudes toward women were shaped by experiences with an aunt and her two daughters who shared the household.

At the age of seventeen, he had his first sexual relationship with a woman when a maid in the home virtually seduced him. He agreed to have intercourse with her on condition that he could bring along a boy friend. After that he had sporadic sexual relationships of the same pattern—two men and a girl. During adolescence, he became infatuated on two occasions with older girls who proved unattainable because of their social position. Feeling deeply rejected and hurt by their aloofness, he had further cause to project his hostility upon women.

Neurotic Interaction in Marriage

The analysis revealed that he was incestuously attached to his sister. On the one hand, he identified all women with the depriving mother; on the other, he was unconsciously identified with the woman. His potency disturbance expressed his unconscious need to punish the woman and revenge himself for the injustices he had suffered at the hands of his mother and aunt; his identification with his sister; and his fear of assuming a masculine role, out of fear of genital loss.

Thus the meaning and structure of potency disorders are seen to bear a relation to the neurotically chosen spouse. In this connection, it may be relevant to mention coitus interruptus, which is withdrawal of the erect penis prior to ejaculation. This practice was once considered by Freud to be the cause of anxiety or "neurasthenia." [7] Today, when the practice is no longer necessary as the only means of contraception, it is occasionally seen as an expression of exaggerated sexual fears. Men who persist in this practice when there is no rational need for it, cultural or religious, are generally expressing neurotic conflict.

On the whole, the prognosis for impotence is more favorable than for frigidity.

Preference for Masturbation

Conflicts derived from childhood masturbation are often carried over into the adult sexual life. Marriage is sometimes used as a defense against masturbation conflict. Not uncommonly, the sexual life of some neurotic adults may be described as masturbatory intercourse. Coitus may be avoided on one pretext or another which really rationalizes unresolved childhood guilt over sexual activity.

An example of this manifestation was provided by a patient who refused to have sexual intercourse with his wife on their wedding night out of fear that he would hurt her. Instead, he stood before her in the bathroom and masturbated into the toilet bowl. Later in their marriage, he would rarely approach his wife sexually, usually waiting for her to seduce him.

The wife was a "masculine" and "dominating" woman who experienced only one orgasm in fifteen years of married life. Her own preference for masturbation was expressed in the mechanics of the sexual act. She would have her husband masturbate her clitoris to the point of climax. Only then would she permit him to insert his penis, thus negating the importance of his organ and establishing her independence of it. At the same time, she was relieving her own conflicts derived from childhood masturbation. She warded off her guilt feelings by making him rather than herself responsible for the act of masturbation, and rejected her own "inferior" genital.

Preference for masturbation is frequently related to the problem of frigidity in that the woman who finds it impossible to attain orgasm through coitus may do so afterwards by masturbating. While not expressing a conscious preference for masturbation, the type of woman who has married a man suffering from potency disturbance often masturbates for relief. Unconsciously, she has created the situation by her neurotic choice of mate.

Men and women who engage in intercourse as part of their marital duty, but attain a higher degree of excitement and satisfaction from masturbating afterwards with fantasies derived from the infantile period are commonly encountered in psychiatric practice. This is true of individuals who are dammed up sexually as a result of neurotic conflicts.

Certain narcissistic types, particularly those suffering from borderline schizophrenia, will frequently lie alongside their marital partners in bed and masturbate, rather than approach the other for intercourse. Such individuals fear emotional intimacy and suffer from extremely ambivalent attitudes in their object relationships. They are withdrawn in their sexual as well as their social activities. Many character types, retaining a primitive level of bisexuality, remain fixated to their perverse fantasies linked to masturbation.

Clinical experience confirms the observation that passive men and masculine-aggressive women often show a preference for masturbation rather than intercourse. Hostile emotions connected with the opposite sex—unresolved penis envy in women and fears of the vagina associated with castration in men—are warded off by resorting to autoerotic fantasies and activities rather than intercourse.

Preference for Perversions

In the perversions, one or more components of infantile sexuality (pregenital impulses) is substituted for genital sexuality. Oral or anal impulses, homosexual activities, voyeurism or exhibitionism, sadism and masochism, and the like replace the more mature heterosexual relationships. Between man and wife the occasional indulgence in some sexual act other than intercourse does not, strictly speaking, constitute a perversion. For example, only when fellatio or cunnilingus is an absolute precondition for intercourse, or replaces intercourse almost entirely as a means of obtaining orgasm, does perversion exist as a clinical entity. Otherwise, such sexual diversions are in the nature of foreplay.

Also in this category are reversals of position, with the woman lying on the man in coitus, acting out a "masculine" role. Such reversal can be considered a manifestation of psychopathology only where it is obligatory for orgastic pleasure.

For example, a young man, in acting out many perverse tendencies on the basis of unresolved infantile bisexuality, preferred to perform cunnilingus on his wife. He rationalized this activity by saying it was the only way he could give his frigid wife some measure of sexual satisfaction. However, he had numerous other indications of polymorphous perverse tendencies, and also suffered from premature ejaculations. He avoided intercourse by making the sexual approach to his wife simultaneously a source of argument. This precluded the consummation of the act. During the course of his marriage, he had several homosexual experiences.

The young man continued a masturbation practice of his childhood in which he would lie face down, spanking himself on the buttocks while rubbing his penis against the bed. In this way he condensed the "crime and punishment" of his second year of life. On occasion he inserted a candle into his rectum. Peeping and exhibitionism were part of his unresolved perverse tendencies. On several occasions he achieved ejaculation by contriving to lean against a woman in the subway. The castration anxiety of his infancy was intense, and was displaced in phobias concerning heights and bridges.

Sado-masochism as a common psychic expression of marital conflict is described elsewhere in this volume. In this chapter, sado-masochism refers to sexual perversion rather than to a social relationship. The difference is seen in the following fragment: A patient complained of the cruel way in which he compulsively mistreated his wife. But he added, "Our sex life together is perfect." His sadism was expressed in hurting, chastising or humiliating his wife, but not in overt sado-masochistic sexual practice.

This frankly perverse sexual structure was revealed during the course of treatment of the daughter of a wealthy man who became enamored of a fellow in a menial occupation because she felt that he and she intuitively understood each other from the start. Early in their courtship they had gone into the woods, disrobed and flagellated each other with willow branches. Each of them found this activity highly exciting.

After they were married, their sado-masochism took a rather

different turn. He openly engaged in extramarital affairs. She would phone the apartment where he was holding a rendezvous just about the time she calculated they were having intercourse. When a voice at the other end answered, she would hang up, feeling that she had interrupted the act. Her husband accused her of making these calls, but she denied it. Violent arguments ensued, and periodically the physical violence became so acute that neighbors had to call the police. This marriage was terminated after two years. Here the sado-masochism was a frank sexual perversion, reproducing the infantile wish to separate the parents in intercourse.

A similar case is that of a respectable, rather effeminate young man, living near a southern resort, who chose a wife with whom he could act out being a slave, a dog on a leash, a beaten boy. On several occasions, he reproduced forgotten scenes of his parents' marital and extramarital intercourse in the following way:

He would have his attractive wife pick up a man in a resort hotel, bring him home and have intercourse with him in his presence. He would then fall to sobbing and ask to be beaten. After such episodes, he would severely upbraid his wife, and call her a prostitute. Feeling deeply hurt, he would be compelled to leave her for varying periods of time. Then would come the inevitable tearful reconciliation, only to be followed by a repetition of the procedure in its entire sado-masochistic cycle.

Extreme sexual submissiveness in a woman is sometimes related in structure to masochistic sexual perversion. Thus, the wife who feels compelled to have intercourse at her husband's request, even though she may be coming down with the grippe or feel exhausted by an unusually hard day, is behaving in a neurotic masochistic fashion.

Such an attitude is akin to a perversion in that the condition for obtaining sexual gratification depends upon some feature in the psychopathology of love. That is, there must be a feeling of one's own insignificance and complete dependence upon the partner, which calls for a compulsive sense of sacrifice or self-immolation in the sexual act.

This kind of ego feeling is reminiscent of the infant's love for the omnipotent parent. Such people have never become complete individuals, daring to have a personality of their own. They need to participate in a union with the "great" partner in order to feel that they exist. Certain of these types have as the unconscious basis of their sexual behavior the fantasy, "I am a part of my partner's body and he can do with me what he wants." *

Not infrequently, these are orally fixated persons who constantly need the reassurance of being loved. Afraid of losing narcissistic supplies, they have not reached the point where they have a mature capacity to love. Their behavior only irritates the partner who feels, "You need me, but you do not love me," and resents this exaggerated and demanding dependence.

The intense masochistic attitude often inherent in extreme sexual submissiveness constitutes a reproach to the partner who has inherited the position of the parent. Extreme sexual submissiveness is also common in men suffering from masochistic dispositions.

Hypersexuality

Most neurotic individuals present obvious and often gross disturbances of their sexuality. They give the impression of being either "oversexed" or "undersexed." Actually, they are dammed up individuals whose struggle with repressed components of infantile sexuality leads either to distorted forms of discharge through pseudo-genital activity or to avoidance of sexual relations.

Fenichel describes the gamut of hypersexual and hyposexual

---

* See Chapter V.

reactions related to the whole range of psychiatric disorders.[4] A psychogenic increase or decrease of sexual desire is not in itself a clinical entity. It is a manifestation of any neurosis: hysteria, obsessional neurosis, phobic disorder, character disorders, psychopathic traits, manic-depressive reactions, and the like.

In cases of hypersexuality there is a neurotic incapacity for satisfaction because the pregenital impulses are constantly being warded off. The Don Juan type, for example, is chronically frustrated in not finding his mother in all the women to whom he makes love. Moreover, in his unconscious life he is frequently impelled by the pregenital aim of incorporation, dominated by narcissistic needs and sadistic impulses which constantly seek expression in his relationships but are blocked by the pathological defense.

The personality of his love objects hardly interests him. What he craves is a negation of inferiority feelings by proof of erotic successes and a bolstering of his masculine feeling by a demonstration of his ability to excite women. The incessant oral-demanding attitude which motivates such types is expressed in the need to make the partner give sexual satisfaction to allay anxiety, guilt or depression connected with infantile sexual strivings.

The capacity for love in this type is like that of a very young infant. He absolutely needs affection, but he cannot yet give it. He is capable of ambivalent feeling and is easily moved to sadistic reactions. In his sexual life there is a condensation of narcissistic and erotic needs. Under these circumstances, sex becomes an addiction rather than the spontaneous culmination of the tendency to love.

The following illustration is a case in point:

The patient is a young married man, age thirty-five, handsome though rather effeminate. By virtue of his business contacts, he had ready access to many attractive women. For years he would have intercourse with four or five different women in one week. He liked to boast of the number of times he was able to repeat the act of intercourse. With his wife, however, he had intercourse only about three or four times a year, and generally on occasions when she was asleep.

He treated his wife as well as his women companions rather sadistically. When one of his sexual partners fell in love with him, he became frightened and asked the therapist to advise him how to get rid of her. One feature of his sexual life—an unconscious feminine identification—is noteworthy: namely, the fact that he wanted the women to describe how they masturbated.

His earliest life was marked by abrupt weaning after his mother became disturbed over her unhappy marriage. His mother would prepare different meals for each of her children, favoring the patient with special delicacies because he was a poor eater.

Early childhood was dominated by fear of a violent father. The patient slept in the parental bedroom, frequently witnessing parental intercourse. As a youngster he bathed with his mother, and experienced erotic sensations as he watched her combing her hair in the bathroom. Because of intense castration anxiety, he unconsciously identified himself with his mother. On one occasion, the boy sought reassurance against this anxiety by displaying his erection before his parents, and was harshly beaten by the father for this act. Shortly after the incident, the boy was seduced by an older brother into submitting to anal intercourse.

As an adolescent, he was quite popular with girls because of his handsome appearance. His preference, however, was for girls who had gone out with his brother. The unresolved oedipal situation was expressed in his compulsive need for an injured third party from whom he could take away the woman.

The infantile element of forbiddance was practically a precondition of his sexuality. He could enjoy sexual relations with another man's wife, or if he felt that he was winning out over his employer by having an affair with the employer's secretary, or was otherwise involved in a triangular situation. However, if the woman became divorced he immediately lost interest.

Hypersexual behavior served to ward off his intense castration anxiety, to ward off passive homosexual tendencies, to restore his self-esteem and overcome depressive tendencies, and to counteract guilt feelings by proving his worthiness of being loved and favored.

The counterpart of this type in women has been previously described in the discussion of nymphomania. The behavior of certain women which appears oversexed is often a reaction formation to an original inhibition and unconsciously serves defensive purposes. The pseudosexual behavior arises not out of sexual needs but out of non-sexual and narcissistic needs. Not infrequently, it is the extended struggle against masturbation and masturbation guilt. Such women may be fixated to an early attitude that sexual intercourse is humiliating or mutilating.

Phobic women in particular are apt to have fantasies of being a prostitute, and in their sexual lives again feel humiliated and revenge themselves upon men in general by symbolically castrating them through sexual domination. Their unconscious vindictive attitudes may be rationalized as a plea for the single standard of sexual freedom.

One such woman suffered from several addictions, in particular to sleeping pills and alcohol. She was markedly hostile to her mother and identified with her father who was a philanderer. She sought out extramarital activities compulsively, chiefly as a counter measure to depression. In intercourse she saw herself as a frightened child being embraced by her mother.

Hypersexual behavior is, therefore, more an indication of anxiety than of spontaneous sexual need. Because of the pleasurable aspect of the symptom, however, individuals so affected rarely come for treatment unless their marriages are seriously threatened, or unless guilt feelings or feelings of dissatisfaction with life move them to seek psychiatric help.

Lack of Sexual Interest

Many people, although they are able to marry and have intercourse, sooner or later develop a disinclination for coitus. Where serious marital discord exists, such withdrawal from sexual activity is understandable on the basis of conscious hostility. However, lack of sexual interest in the course of an otherwise stable marriage indicates a neurotic development based on the reactivation of specific infantile conflicts.

Depressive men and phobic women, for example, are most apt to show lack of sexual interest in the exacerbation of their neuroses. Usually, there are concurrent neurotic manifestations, such as work inhibitions, travel phobias, fear of enclosed spaces, and the like. These are displaced derivatives of the fear of sexual and aggressive excitements.

Many men of this type have unconsciously identified their wives with their mothers, particularly after the wife has given birth to a child, and has thus actually become a mother. This circumstance activates guilt connected with oedipal fantasies, which is warded off by the defensive measure of repressed sexual desire.

In other cases, aggressive or sadistic fantasies have been activated in some way, resulting in the cessation of sexual activity. For instance, an obsessional neurotic man read in a tabloid newspaper of a woman found murdered by a sex maniac who had shoved a corncob into her rectum. On reading the account, the patient developed marked anxiety and began to be tortured by doubts as to whether he had not done the same to his wife. For months he made no sexual approach to her.

This symptom proved to be a defense against his own repressed wish for and fear of anal penetration by his father. The analysis of his wife had revealed an interesting corresponding fantasy. She unconsciously had a fear of and wish for anal impregnation. She, too, was disinclined to have sexual intercourse. This was a unique situation in that there was an opportunity to study the dovetailing fantasy life of both partners in a neurotic interaction.

The following illustration is presented in more detail.

A woman suffering from obsessive feelings of inferiority and insomnia, expressed marked lack of interest in sexual relations after the birth of her second child. She would often sit up until two or three in the morning to avoid having intercourse. She stated that early in her marriage, some twelve years before, she frequently attained orgasm. When her husband returned from army service after the war, the couple had gone to live in her parents' home. Out of anxiety that they might be overheard having intercourse, she began to be unresponsive in the sexual act. But when they moved out of the home, her sexual revulsion became even more marked, and on the rare occasions when they did have intercourse, she failed to attain orgasm.

Actually, her sexual difficulties were aggravated by the birth of a child with a withered arm. The patient then became extremely tense and nervous and began to suffer tormenting doubts about "what people will think of me." Shortly thereafter, she developed transitory subway phobias, cardiac palpitations, insomnia, indecisiveness and exaggeration of her compulsiveness in regard to housekeeping.

Her early life was marked by fear of her father who became abusive when he drank excessively. She was dominated by an older brother. Her mother was strict, fanatically religious and domineering, and the patient remained dependent on her even into adulthood for minor as well as major decisions.

Two traumatic episodes were related to her deep sexual guilt and inhibitions. At the age of six, she was seduced by an uncle who had her masturbate him. At the age of ten, she had a similar experience when the owner of a bicycle shop, ostensibly teaching her to ride, masturbated her. The birth of her deformed child, a visible symbol of her "sin," mobilized her guilt regarding sexual activity in general, especially its primary incestuous significance, and her anxious infantile fantasies of punishment by castration. Part of her defense against intense anxiety was withdrawal from sexual activity. In her case, there was inhibition of social and sexual interest, as had happened before in her life between the ages of six and sixteen.

Several other types of neuroses illustrate the variety of dynamics responsible for pronounced lack of sexual interest: The man avoids coitus because of his repressed fantasy that the vagina has teeth. A husband loses sexual interest in his wife when she has given up her job and thereby becomes "castrated," unconsciously reactivating his own castration anxiety.

The woman whose husband has acquired the unconscious significance of her mother has her hostility activated thereby. She avenges herself on her mother in effigy by punishing her husband through the avoidance of sexual contact.

Then there is the woman whose defenses against unconscious homosexuality were intensified after she learned that her husband had had an affair. Although she forgave him, she lost interest in sexual activity for almost a year.

Another instance is that of a man who in childhood sought to escape from fantasied mutilation at the hands of his father, and through symbolic self-castration, had taken refuge in identifying with his mother. Passive sexual attitudes and aims were evident in his material. While marriage served as a defense for him against his feminine identification, he rarely desired intercourse. On those few occasions when he did perform coitus, he had to have the stimulation of hearing from his wife the minute details of her sexual life with other men.

Certain similar types ward off the feminine identification through projection. That is, they accuse their wives unjustly of sexual activity with other men. Their paranoid attitude means, in effect, "You are sexually interested in men, not I."

Summary

The prevalence of sexual inhibitions and symptoms in marriage is not remarkable in view of the fact that most people suffer from psychoneuroses of lesser or greater degree. Psychosexual disorders may occur in the context of either a stable or an unstable union. While sexual conflicts are usually features of discordant marriages, they are not at all uncommon in manifestly harmonious relationships. Paradoxically, there are certain sexual difficulties which exist because of love for the partner, based on a dissociation of tender and sexual components.

The psychosexual problems most frequently encountered in marriage are those of frigidity, impotence, preference for masturbation, preference for perversion, hypersexuality and lack of sexual interest. These are all derived from the unconscious infantile instinctual conflicts of one or both partners, and are activated by some aspect of the relationship between the mates.

Proper psychiatric treatment is directed not to the sexual manifestation itself, but to the underlying neurosis. The prognosis depends upon the clinical diagnosis and the psychodynamic evaluation of the specific case. Factors to be considered are the age, motivation and circumstances of the patient. Drug or hormone therapy is usually of no avail in such cases. Certainly, advice on the techniques of coitus, as prescribed in the marriage manuals, can do very little to remedy such disorders. In some mild cases of sexual inhibition due to situational anxieties, casework treatment may result in improving the sexual adjustment of a marriage.

Because sexual disorders are deeply rooted features of neurosis, only a well-planned and sensitive therapeutic approach can effectively modify the disturbed psychosexual attitudes.

References

1. Bergler, Edmund, "The Problem of Frigidity." *The Psychiatric Quarterly*, XVIII, 374, 1944.

2. Eisenstein, V. W., "Dreams Following Intercourse." *The Psychoanalytic Quarterly*, XVIII, 154-172, 1949.

3. *Ibid.*, p. 168.

4. Fenichel, Otto, *The Psychoanalytic Theory of Neurosis*. New York, W. W. Norton & Company, Inc., 1945.

5. Freud, Sigmund, "Contributions to the Psychology of Love: The Most Prevalent Form of Degradation in Erotic Life." *Collected Papers*, Vol. IV. London, Hogarth Press, 1934.

6. ———, *New Introductory Lectures on Psychoanalysis*. New York, W. W. Norton & Co., Inc., 1933.

7. ———, "The Justification for Detaching From Neurasthenia a Particular Syndrome: The Anxiety-Neurosis." *Collected Papers*, Vol. I. London, Hogarth Press, 1924.

8. Kinsey, A. C., Pomeroy, W. B., Martin, C. E. and Gebhard, P. H., *Sexual Behaviour in the Human Female*. Philadelphia, W. B. Saunders Co., 1953.

Interaction *Between* Psychotic Partners

———————•◦•———————

I   Manic-Depressive Partners

EDITH JACOBSON, M.D.*

AS A PRACTICING PSYCHOANALYST, my field of observation with regard to psychotics is naturally limited, and my experiences with them have been intensive rather than extensive. I have, however, had an opportunity to study marital situations in couples where one or both partners has been manic-depressive, as well as a few who were frankly schizophrenic and, while more clinical confirmation is needed, have noted the differences in object choice and interaction of these two groups of psychotics.

My findings regarding the object choice, the personal relationship, and the pathological interplay of manic-depressives and their partners may be described as follows: The marital relations of such couples during free intervals often appear to be very good and warm, and they may be at a rather high personal level, as I have stated in previous papers.[1,2] As long as they are not sick, manic-depressives can be delightful companions, as pointed out by the elder Bleuler and other old-time clinicians.

---

* Instructor, New York Psychoanalytic Institute.

Such patients may show emotional warmth, sexual responsiveness, a capacity for rich sublimations—in short, features which greatly attract their partners. However, if we examine the marital relations in such cases more closely, we can detect the pathological elements which harbor the germ of future breakdowns of the sick partner. We find that the seemingly over-close, over-warm marital relationship is actually of a symbiotic type. The partners greatly depend and lean on each other, cling to one another, actually feeding on each other.

The most favorable constellation seems to be in couples where this mutual dependence has a sound basis either in practical collaboration, as for instance in professional work, or in a sharing of interests and hobbies. But in many cases the apparent over-closeness and mutual over-dependence of the couple is combined with a striking lack of real interests in common. The partners are in great need of each other, but have nothing to say to each other, even though each may have a rich, resourceful ego and actually have a great deal to talk about. This situation promotes a type of marital conflict that is bound to precipitate depressive periods in the manic-depressive partner.

When we have the opportunity to investigate the personality structure of either partner more thoroughly from the psychoanalytic point of view, we find that both, as a rule, are pregenitally fixated, usually oral character types though often of quite different variety. Thus a sort of oral interplay will develop in the couple which may build up mutual increasing demands to the point of inevitable disappointment and breakdown in one or the other partner. Sometimes both partners are manic-depressives who may have to be hospitalized alternately. In one such case the husband went into a hypomanic state after his wife's suicide.

In another instance, a man who had always been gay, good-natured, easy-going but rather selfish, married a girl who for twelve years had had depressive episodes each year. These had spontaneously subsided after her father's death, whereupon she fell in love with her future husband. They had an affair for two years, during which time she had no real depression. After their marriage the girl remained in comparative balance, vacillating between mildly hypomanic, overactive states and brief periods of mild depression. But her constant domination and pushing of her more passive husband resulted in his becoming a chronic depressive with incessant hypochondriacal complaints.

In other cases the partner stands up amazingly well during the depressive period of the husband or wife, but goes into depression right after the other's recovery.

As to more specific features regarding the object choice and oral interplay in such couples, the fact is that many depressives are very masochistic in their choice of partners. As long as they are not sick, they tend to ward off their own infantile oral demands by playing the active, self-sacrificing but also domineering mother. In return for their practical support they expect the moral support they need so badly for maintaining their self-esteem. In other words, to the manic-depressive partner the love object represents above all a super-ego figure; that is, a highly overvalued love object in which all his love is invested, for which he is ready to sacrifice everything in the hope of getting love and praise.

Indeed, the partner is to the manic-depressive the medium through which he lives and which he needs for his mental balance. Frequently, however, we see that the so-called healthy partner is actually the more passive, more selfish, more frankly demanding character type. He is glad to repay the services rendered him by expressing his deep gratitude and praising his partner—an attitude which is often not backed up by his practical behavior.

Of course, there may be all kinds of combinations in the division of the mother and child roles, such as where one partner supplies money and material goods, the other intellectual and spiritual nutrition and the like. More often than not, however, the manic-depressive partner, male or female, when in a healthy state, seems to be more active, strong, supporting in a material and practical sense, while depending on moral support from the other by way of continuous love, praise, recognition and encouragement. This precarious balance is naturally apt to be easily upset. Frequently, the neurotic partner, having tired of his own role in this barter agreement, takes more and more advantage of the other's masochistic attitudes and becomes increasingly dependent and demanding. This is what appears to precipitate the depressive conflict in the other partner. Thus, in cases where both partners can be studied, it is not surprising to learn that the patient's complaints about the love object, which underlie the depressive self-accusations, hit precisely at the core of the marital problem.

For example, a manic-depressive woman, a social worker, blamed her husband for being childish, selfish in a primitive way, though he pretended to be an idealist and to share her political and social beliefs. It was her opinion that her husband clung to her merely because he was weak and passive; that he had chosen her only because he admired her higher social background, her better education and her earning capacity. She felt sure that now that he had obtained a well-paying job himself, he would like to get rid of her so that he could marry some simple, stupid girl who would constantly serve and cook for him. The husband denied all this. He made a great showing of concern for her, but actually failed completely when it came to bolstering her during her depressive periods.

The woman suffered a relapse at a time when I was unable to take care of her. Transferred to another psychiatrist, she was given shock treatment, and a week after being discharged from the hospital she committed suicide. The husband immediately went into a brief hypomanic state. After he had sobered down, he came to consult me about some sexual difficulties. It was at this time that he confessed he had never really cared for his wife. He admitted having chosen her because she was so superior to him. While admiring her genuinely, he felt unable to love her and live up to her high standards. Rather brazenly, he stated he was glad to be rid of her.

Right after her death he moved to the city, boarded their child with some friends, and took to consorting with prostitutes. After calming down, he became acquainted with a simple girl who was a homebody with no intellectual interests. She would take care of him and this was what he needed. He did marry this girl, and they got along well.

The following is a similar case: Some years ago, a man came to see me whose wife I had treated in 1928 for a severe depression. At that time she, too, had complained that her husband did not really love her, since he would flirt with any pretty girl who happened to be around; that he was weak, inconsiderate and tactless; that he was superficial, did not care for their baby, was immature, did not want to assume the role of a father, etc. Subsequently, during another depression, this girl committed suicide, as had her father and grandfather.

Twenty years later, when I met the husband in New York, he confirmed every one of his wife's accusations. He had married her mainly for her money, had felt tied down and burdened by his early marriage and fatherhood, had wanted to be free and to continue his bachelor life. His affection for his wife had faded from the moment she became pregnant. Unlike the husband in the first case whose second choice was much sounder, years later this man married a woman who promptly also developed depressive periods after the wedding. The same pattern repeated itself a third time with a girl friend.

In another case, the husband of the depressive patient went into analysis during the final stage of her treatment. His analyst, unaware of the complaints of the wife which had become conscious during her treatment with me, gave me a gloomy picture of the man's infantile selfishness, of his demanding possessiveness toward his wife, coupled with his feeling of being entitled

to cheat her as much as he desired. All these features corre-
sponded precisely with my patient's description of her partner.

In all of the foregoing cases, the deceptive over-dependence
of the partners on their wives made it difficult for an outsider to
believe the patients. In the last mentioned instance, an amazingly successful treatment of the husband by his therapist has
achieved excellent and lasting results. The consequent improvement of this marital relationship has so far prevented my patient
from any further relapses.

Thus my experience has shown that however exaggerated
the patients' hurt, disappointment and hostile derogation of their
partners may be, their complaints are usually more justified than
may appear on the surface.

As to the marital situation as it may develop during the
manic-depressive partner's periods of sickness, in many cases a
rebellious stage precedes the depression proper. During this period the patient often frankly disowns the role of the active, supportive mother or father toward the partner. Several patients
have voiced the same typical remarks: "He is such a baby, so
selfish, but what about me? I cannot always mother him. I need
a mother myself." Thus we may say that the patient's depression
aims at forcing the partner to take over the maternal role.

The retardation, the helplessness, the exhibition of self-dero-
gation play on the partner's pity, his sympathy, his guilt feelings.
Of course, we are all familiar with the special quality of the
hidden sadism in the melancholic, which everyone finds so hard
to tolerate. Even in cases where the aforementioned rebellion
and complaint about the partner do not become manifest, the
latter invariably senses the reproaches and the hostile appeal
underneath the patient's self-punitive attitudes.

In other words, the depressive patient never fails to make his
partner, often his whole environment and especially his children
feel terribly guilty, and to pull them down into a more and more
depressed state too. This explains why the supposedly healthy

partner, in defense, so often becomes amazingly aggressive and even cruel toward the patient, and may hit him precisely on the most vulnerable spot.

To illustrate, a physician, whose wife had gone through a series of depressive periods after her menopause, appeared to be very understanding in discussing his family situation with me. He knew full well that his wife responded to any blame or demands with an increase of her depression. He wanted to be patient, to consider her pathological retardation and reduce his claims. But he could not control his tendency to make continual demands and to accuse her of little failures, particularly during her periods of depression.

One day, on their way home from a party, he suddenly complained of feeling ill. Seeing that he was close to fainting, the patient suggested he let the car be picked up by the garage so that she might get him to bed quickly. Immediately he turned on her, saying, "You are responsible for my illness. You have made me sick—you are killing me!" Whereupon his wife, who had been improving, promptly had a relapse.

Such scenes are not a rare occurrence among couples where one partner is severely depressed. When the depression lasts long enough and the patient is not removed from the family, the healthy partner slowly succumbs to his companion's severe hidden hostility and tries to ward off his own depressive reaction by aggressive counter-accusations which intensify the patient's pathological feelings of worthlessness. In this way a vicious cycle is set in motion.

When the partner tries to escape from the patient's depressing company by looking for comfort on the outside, either in work or in social activities or in a sexual affair, the patient's feeling of being unloved is bound to increase. When the depression of the patient lasts long enough, the other members of the family will nearly always become infected, as it were, and eventually go into a reactive depression themselves. Briefly, with regard to the effect of such conditions on the children, sometimes the most touching situations may develop, with the children trying to elicit even a smile from the sick parent, or doing what little they can to make the parents love each other and be happy.

As to the hypomanic states of the patient, in one case I saw a husband, who had been relieved of the terrible strain of his wife's depression, go into a mildly elated state during which he was constantly calling me up to assure me how wonderfully everything was working out.

In other cases the partner finds the patient's hypomanic state even less tolerable than the depressive state, especially if the patient is aggressive—a constant nuisance as it were—without awareness of his aggression and in general without insight. Where the hypomanic or manic state leads to sexual escapades or to a careless spending of money, the partner may react with shock, disgust or frank fear and finally may himself go into a depression. Such responses may lead to an ultimate disintegration of the marital relationship, which the depressive phases had not precipitated.

It should be emphasized that the partly justified reactions of husband or wife to the partner's acute phases of illness, reactions which have a pathological quality in themselves, inevitably increase the patient's conflict and that a vicious cycle arises in the pathological interplay of these couples. I mentioned that most often the other members of the family, especially the children, become deeply involved in it. Thus a generally sick family situation may keep developing up to the point where the patient withdraws more or less completely from the whole family. This is one of the reasons why clinicians have recommended hospitalization or removal of the patient from the family environment and only rare visits by the family, at least during the severest stage of the psychotic depression.

However, there are certain patients who react very badly to

hospitalization, and make a much better recovery when brought back home. This may be particularly true of patients who struggle very hard not to withdraw their libido from their love objects completely, and who therefore feel that the very presence of the partner helps to maintain the link, however ambivalent it may be.

Finally, I should like to comment on the "contrast" marriages, i.e. between manic-depressive and schizophrenic partners. In those few cases I remember, I have never seen a marriage between a manic-depressive and a schizoid type work out. The manic-depressive invariably broke down because what he really needs is warmth and affection.

On the basis of their individual emotional needs, it is difficult to see why a manic-depressive would want to marry a schizophrenic. In one such case, the husband was a manic-depressive who usually showed a mixture of compulsive and mildly hypomanic attitudes. In this state he was often very active, very loving, very much out to serve and give love to his partner. Some of these types may perhaps attract schizoid people whom they "warm up." However, I have never seen a stable equilibrium between such contrasting types.

Regarding marriages between schizophrenics, discussed in the following section, I have observed three characteristic types of object choice in these psychotic relationships. Those who are able to have some sort of personal relations beyond temporary sexual affairs, may tend to integrate at a lower level by marrying partners who are socially and intellectually their inferiors. Others marry schizophrenic partners with whom they live in a common unrealistic world.

A third category are schizophrenics who attach themselves to rigid, compulsive partners. The compulsive reaction formations of the latter seem to serve the special function of strengthening their own compulsive defenses against the threat of a breakdown. While this type is more frequent, in some cases I have seen the very compulsiveness of the partner precipitate an episode because it could no longer be tolerated. Such compulsiveness is frequently seen in latently melancholic individuals.

References

1. Jacobson, Edith, "Primary and Secondary Symptom Formation in Endogenous Depression." Read before the American Psychoanalytic Association. New York, December 16, 1947.

2. ———, "Contribution to the Metapsychology of Cyclothmic Depression." *Affective Disorders*. Phyllis Greenacre, ed. New York, International Universities Press, 1953.

II Schizophrenic Partners

GUSTAV BYCHOWSKI, M.D.*

MY CONCLUSIONS on the pathology of interpersonal relations are based on cases observed in private practice as well as hospitalized patients. The former were mostly latent schizophrenics, the latter active, overt psychotics.

For purposes of clarity, this study is organized along the following basic lines, some of which obviously overlap:

1. The motives for the choice of the marital partner.
2. The pattern of relationship which obtains between the partners.
3. The course of marriage.
4. The influence of therapy on the course of marriage.

There can be no doubt that the choice of a psychotic partner is a symptom as such, added to the already existent psychopathology and complicating the clinical picture. Further, the primary and secondary gain which the psychotic partners derive from the marriage must be an important reason for their maintaining a marriage which to an unbiased observer may appear devoid of any meaning, fraught with frustration, and even rather horrible in some ways.

The following clinical illustrations are gleaned from the hospitalized cases I have observed:

The husband was a pleasant man, thirty years old, of slight build and very limited intelligence. He was never able to get through grammar school, attended ungraded classes and failed them. Psychological examination revealed mental deficiency and schizophrenic deterioration. At the time of his hospitalization no other psychotic symptomatology could be observed.

---

* Department of Psychiatry, New York University College of Medicine.

On psychiatric examination he disclosed that he never liked to study and did not know very much. "My wife," he said, "helps me with information. She knows much more than I. She helps me to get along." He felt run down and had come to the hospital to find out just how run down he was and how much weight he had lost. "It never hurts to find out," he said. Yes, he wanted to work—after all, he had a child and was married.

In contrast to his wife who, as will be seen, complained in a paranoid and agitated way about the various places where she, her husband and the child found shelter, Mr. B. was placid, smiling, almost contented. The problem of his future and that of his family did not preoccupy him. Though he had not worked in the past, he felt that somehow he would find work in some indefinite future.

Mrs. B. was twenty-eight years old. She used agitation and paranoid accusations as a means of covering up and denying her phobias, anxiety, depression and a sense of threat by an impending catastrophe. On examination, when asked why she had married her husband, she explained that she had felt sorry for him—he was so lonely and there was nobody to take care of him. "We both needed each other," she said.

She stated that she had been warned by her brother-in-law before the marriage that her husband would never work. She wanted to hammer it into him that it would be easier for him to find work if he dressed well. She never thought of divorcing him, though she had been advised by her relatives to do so.

Mrs. B. wanted to make a go of this marriage. The fact that she became pregnant right after the wedding proved to be a real misfortune, and her relatives gave her no help. She had lost her mother when she was still a child. Her father, who was very poor, lived with her sister. After graduating from high school, the patient had gone to work and lived with another married sister. Her ideal of a man was a "nice fellow," who would respect her and not make passes right away. So many of the men she met were crude and tough.

Mrs. B. was full of accusations against other patients with whom she fought. She felt indignant toward the Welfare Department because of the poor homes that were assigned to the family. The Family Shelter reported that they found the couple irresponsible, lacking in judgment and demanding. Mrs. B. gave the staff members a difficult time. She had outbursts of bad temper, was aggressive and assaultive and displayed paranoid attitudes and unrealistic ideas as to what the Shelter should do for her.

The child was rejected and neglected by both parents, each of whom wanted to shift his care onto the other.

The wife's excuse and alibi for her husband's unemployment was, "My husband is crazy and cannot work." He was completely dependent on his wife and rarely answered questions in her presence. In fact, she took over and answered for him. At times, she threatened to leave the child and her husband, feeling that they were preventing her from making her own adjustment.

In this marriage the question of the choice of mate presents no problem insofar as the husband is concerned. In his childish immaturity, aware of his intellectual limitations, he leaned on her and demanded her help. Thus to him she was an obviously maternal, perhaps even paternal substitute. She supplemented his deficient intelligence and his lack of aggression.

On the other hand, the wife's choice was based on more complex motivations. Her husband's general inhibitions coincided with the pattern of her super-ego. She liked him because he was a "nice boy" who respected her instead of treating her like an object of sexual desire. Moreover, he appealed to her conscious wish to mother and help him, as well as to her unconscious desire to direct and dominate him. The marriage was intended to reverse her own life situation in which she had no home of her own and was dominated by her married sister, while her father was appropriated by another daughter.

Her own dependency needs and unresolved primitive aggression became apparent in her paranoid-aggressive behavior. In directing unreasonable claims and complaints against various institutions, she was treating them as though they were her own unsatisfactory family. The unconscious aggressive component of her maternal attitude toward her husband became apparent in her neglect and rejection of her child.

A further determinant in her marital choice may have been the similarity between the personality of the husband and that of her father—both weak and ineffectual. Without deeper analysis, we can only assume the masochistic character of her choice, since she had been warned that her husband could not work and support a family.

What made this choice specific and pathological in its outcome is not merely its determination by unconscious motives, but the fact that the ego submitted to them with an utter disregard of reality. The same holds true for the husband's choice of his partner, except that, owing to his limitations, his role was a rather passive one. In view of the dynamic constellation of both partners, this marriage necessarily disintegrated into a pattern of disorganization, with welfare institutions playing the role of inadequate family substitutes.

In some schizophrenic couples, a curious pattern of mutual adaptation occurs which allows for a temporary and relative equilibrium. Usually, one of the partners adapts himself to the other and takes over some of his psychopathology by identification. This type of conjugal psychosis may be considered a special form of the *folie à deux* described in the literature. In other cases, however, such a temporary solution proves impossible. This may be due to various causes, mainly to what may be called the incompatibility of the pathological personality structure, when the psychopathology of one partner is such that no measure of adaptation can secure for the other any semblance of equilibrium.

The following hospital records will illustrate the type of interaction observed in severely psychotic couples.

Mrs. L., a nineteen-year-old Negress, had come to the hospital for help. She was withdrawn and depressed. Her husband, she said, thought she was unfaithful because there was some menstrual blood on her sheets. He accused her of having relations with other men, even when he was gone for only ten minutes.

The couple had beaten their first child so severely that it had died in a hospital of a subdural haematoma. Now her husband wanted her to help him punish their second child. The patient spoke in symbols and allegories, with religious overtones. She accused herself of having hurt her second child. She said her husband beat her, never let her out alone, and did not allow her to look into a mirror or out the window. "I should have run away before," she said. "I don't know why I didn't." After a minor argument, she had left her husband and baby and refused to go back.

Mr. L., age twenty-nine, was evasive and refused to answer questions. "I have been over all this before," he said. "Why don't you read the chart or talk to my guardian?" While visiting his wife, he too had been admitted to the hospital. He denied all his wife's accusations, and his answers were paranoid. "My wife and I have difficulties. She hides things from me. She had something under her sweater. She would not tell me what. It was a rag of blood. I came here to prove that I am a good husband, not lying as she says I am."

He had been discharged from the Army with 100 per cent disability after having some shock treatment. He was unemployed and played basketball most of the time. His superficial façade covered up a deep-seated aggression which threatened to break out from time to time in the form of impulsive behavior. The psychological examination revealed a chronic schizophrenic who, while not completely alienated from the world, was frankly psychotic.

At times, the similarity of the pathological structure accounts for the marital choice, especially if both partners come from marginal economic and cultural groups of society. In these cases the choice is both based on and favors further reciprocal identification. The wife in the following illustration was observed in the hospital; her husband was interviewed from time to time.

Mrs. R., age twenty-one, had been admitted after attempting to commit suicide by taking a large dose of barbiturates. Her father, to whom she was deeply attached, had died when she was a child, and she had lived with her mother and older sister. She was very unhappy at home, with constant friction between herself, her sister and her mother. She felt quite inadequate at school, failed in her classes and left without having been graduated at the age of fifteen. She tried to work but could not hold on to a job, being overly sensitive to criticism, hurt by the slightest remark and very shy.

When she was sixteen years old, she met her future husband who was then nineteen. She married him because she became pregnant, and to get away from home. They went to the west coast where the husband had some friends who helped them financially and to make

arrangements to have the child put out for adoption. For years Mrs.
R. remained inactive, neither working nor taking care of the household. The couple had had no sexual relations with each other for
several years, though they slept in the same bed. However, the wife
was promiscuous with many other men, hoping to find a protector
and helper.

Mrs. R. had a feeling of futility, depression, apathy and emotional detachment. She had never experienced sexual gratification
with any of her lovers. Upon testing, her intelligence proved to be
good, but her judgment was poor and she had no conception of planning for the future. She had never thought of divorcing her husband, though she neither loved nor respected him.

Mr. R., twenty-four years old, was a college graduate and had at
one time done postgraduate work in psychology. However, working,
studying and taking care of his wife obviously proved too much for
him. In order to support her and himself, he resorted to taking a job
as a cabinet-maker in a factory. He was retiring and shy, and had
been undergoing psychotherapy for some time. He had never had
sexual relations with anyone but his wife and even this had ceased
for the past three years. He described it as a "lack of affect" in
himself. When asked why he permitted his wife to sleep with other
men, he replied, "If it's good for her, that's fine."

He had married her because she became pregnant although he
had realized at the time that she was sick. He had a feeling of responsibility toward her, however, and wanted to take care of her.
He would not take any active part in divorcing her or changing the
situation.

The attitude of passive clinging to a completely inadequate
partner becomes particularly striking when it is the woman who
clings to a deeply disturbed, schizophrenic husband, and it is
indicative of serious pathology in the wife as well. In her personality structure we find elements of masochism, deep feelings
of inferiority and, at times, deep-seated hostility directed at a
parental figure. In maintaining her attachment to an inadequate
husband, she unconsciously aims at hurting her parents.

The following case, treated by a colleague,* is a striking illustration of such a constellation.

---

* Dr. Bernard Goertzel.

Mrs. S., a thirty-one year-old white woman and an artist of considerable talent, married a Japanese artist who was suffering from a manifest schizophrenic psychosis. She stayed with him for ten months, though he never had sexual relations with her, never worked and let himself be supported by her. He would leave her without notice and then come back just as abruptly.

Asked why she had married him, Mrs. S. explained that she did not feel "worth anything better." She could not divorce him out of apprehension as to what would become of him. Besides, she was afraid to be alone. Her attachment to him covered up her hostility, since she admitted to a compulsive desire to hit him.

She felt that she had never lived up to the ideals of her father, of whom she had built up a wholly imaginary picture. In reality, he was a small, shy, ineffectual man who attempted time and again without success to accomplish something of real importance. Her father had forced her into many things, including attending college. As an expression of rebellion, she had quit before graduating.

Mrs. S. felt a deep fear of the outside world which appeared to her to be fraught with danger. Afraid of men, that is, of the dangerous father figure, she found pseudo-security in a marriage to a weak man where she could play the masculine role. By avoiding sexual relations, she was able to protect her latent homosexual attachment to an idealized mother.

After their divorce, which was made possible by analytic therapy, she remained for months in a state of diffuse depression with many schizophrenic features. Projective techniques have helped to outline her personality structure and correlate it with her marital choice. Manifold reactions of denial served to cover up a core of deep ego weakness. Playing the leading part in the marriage helped her to compensate for this ego weakness by a false image of a strong self and a façade of grandiose exhibitionism. In this way, she could act out a fictitious ego ideal shaped according to her father's wishes, while at the same time punishing herself for not really living up to her father's ideals.

As the following case will demonstrate, the complex structure of marriage between two schizophrenic partners reveals itself only in psychoanalytic investigation. I observed the husband; the wife was seen by a colleague.*

---

* Dr. Edward Joseph.

Mrs. X., a twenty-two year-old divorced woman, first came to psychiatric attention when she was hospitalized at the age of seventeen. She presented a classical picture of anorexia nervosa of about one year's duration. Her symptoms were typical—anorexia (which earlier had alternated with bulimia), constipation and amenorrhea which antedated her actual loss of weight. Although she was extremely active and showed no evidence of real malnutrition, she weighed only seventy-two pounds at the time of admission, and lost six more before it was possible to help her overcome her eating inhibition. Later on, she achieved a weight of about 110 pounds, and has maintained it at that level for the past several years.

The patient showed the usual fears and fantasies involving oral impregnation and an identification of food with the impregnating phallus. The type of object relations she developed over the years was significant. As the only child of a Jewish father and Catholic mother, she herself had no formal religious training, and alternated between regarding herself as Jewish and Protestant. This confusion about her religion was characteristic of her attitude toward herself in many other respects. Her parents also alternated in their attitudes toward her. Her mother, usually strict, would suddenly become yielding when the patient wanted her to be rigid. On the other hand, the father, generally yielding, had unexpected moments of sternness.

The mother was often violent in her expressions of rage and there were many scenes between the two women, often provoked by the patient, in which the mother would strike or beat her. On other occasions the patient would barricade herself in the bathroom, emerging only after having extracted some sort of concession from the mother. In earlier years these quarrels were about various matters, but from adolescence on they either had to do with sexual activity or with food.

The patient had very few friends in early childhood, and after the onset of puberty she even withdrew from these. Once her anorexia began, however, she cultivated many platonic relationships with boys, always breaking them off if they threatened to develop further. To compensate for her lack of dates, she took to eating quantities of the fruit of the same name, so that at one period dates were the only food she consumed. She herself made this symbolic connection. Similarly, at a later period when she felt lonely for her one girl friend named Cookie, she found herself eating dozens of cookies. In spite of a conscious attitude of being very considerate of others, she actually had no awareness of other people's feelings and by and large treated them as objects which were either useful to her or not, depending on the circumstances.

When she married a Jew she thought she would attempt to be Jewish for his sake. On the contrary, she hated his family and was contemptuous of him in all matters concerning his religion. The numerous scenes with her mother were repeated with her husband, and here, too, she used food as a bone of contention. There was a constant denial of tender feelings toward him, of her own sexual desires and of aggressive feelings, except for reactions to his anger. The idea that it was she who provoked many of these scenes, as she had done earlier in her life, was unacceptable to her. After the birth of a son, she regarded the child as her property, treating him at best as well as she treated her dog. Later on, established a relationship with a Catholic man which was similar to the one she had had with her Jewish husband.

Her transference attitudes in therapy were characteristic. There was bland denial of any feeling toward the therapist, and when confronted with evidence to the contrary she insisted she had to feel neutral about him or he could not be of use to her. Rather than face a heightened transference of either positive or negative nature, she would break off therapy from time to time, and also obstructed any attempt at analysis.

Turning now to the husband, whom I treated, an analysis of his marital choice and of the function of marriage in his psychopathology is as follows:

When Mr. X. came to me for help he was twenty-five years old, working in his father's business, still married but living alone. He appeared to be a pleasant young man with a superficially good rapport. Yet his empty smile, inappropriate laughter and communication limited to a few monosyllables (Bleuler's "short associations") were discouraging.

My questions, he said, made him feel like an insect on a pin. He was afraid he was a homosexual, that he was turning into a woman. Panicky and "terribly insecure," he was unable to convey any precise image of his parents. He was staying at a hotel in the hope that he would be able to bring a girl up to his room. However, he was afraid to approach anybody. Though he denied any delusions or hallucinations, the clinical impression was that of a latent psychosis in the process of assuming an active course. This impression was soon confirmed by the early analytical sessions as well as by the results of psychological testing.

Attempts on the part of my patient to resume life with his wife gave new evidence of Mrs. X.'s pathology as a marital partner. Yet he continued to cling to her for many months. It was only as a result of considerable analytical work that he was finally able to relinquish and divorce her.

His analysis disclosed a considerable degree of fragmentation of his ego, with large sectors functioning on a primitive infantile level. Here all the conflicts of his early childhood remained unresolved and, as in the past, he felt tied to his parents and older sister by dual bonds of intense dependence and violent hostility. Fear of being abandoned by one or all of them remained as intense as in times gone by. During his marriage ceremony he actually sobbed at the prospect of leaving his father.

In this archaic part of his ego he was unsure of his own personal identity, since he still felt himself so deeply rooted at first in his mother and then in his father. He continued to preserve his full primary narcissism with its megalomania and unsatiable desire for absolute love and protection. Frustrations of his pre-oedipal period had laid a solid foundation for his oedipal conflicts. One of the solutions he chose was complete identification with his mother, with the resulting passive feminine attitude toward his father. This inverted oedipal constellation was enhanced by homosexual trauma which he had suffered at the age of seven.

Another element in his disturbance was the hostile identification with his father in the oedipal conflict and the preservation of incestuous wishes directed toward both mother and sister. All these primary constellations remained unaltered, resulting in a defensive splitting of the ego. The all-giving mother was as alive within his ego as the rejecting phallic mother image. Similarly, the incorporated image of the benevolent and omnipotent father coexisted with that of the irate rival and castrator.

Why had this patient clung to his wife, and why had he married her in the first place? To him the most natural though superficial explanation was that she was the only child of wealthy parents. Yet this was a strange motive in view of the fact that he himself came from a well-to-do family and had a good and sufficiently independent job in his father's business. Actually, his interest in his bride's wealth was an indication of his overwhelming wish for support and passivity. In his passive mood he was reliving the fears of his early childhood when he was afraid of being abandoned by his parents. He hoped that when he married, his wife and her parents would become convenient parent-substitutes, more reliable than his own parents because as yet cathected with little ambivalence and no feeling of guilt.

Mr. X. felt lonely, isolated from his parents and from the rest of the world; he longed in vain for love and support, deeply unsure of his masculinity and beset by passive feminine urges. His young bride seemed strong and self-sufficient. Her apparent coldness attracted him. Not only was it a challenge, but he also felt that once he "would get into her" they would become a closed circle—he would have no competitors, nobody would interfere with him. "Since she seemed so strong," he related, "I could substitute her for my penis. I could smash myself against her by conquering and raping her."

Moreover, he was attracted by her frigidity and her obvious revulsion toward the sexual act because he, too, felt with a part of his ego that "sexual intercourse lowers the human being." Since sexual intercourse was something to be ashamed of, his relations with her were a repetition of his sexual experience with his sister and a girl cousin in his early adolescence. The analogy went so far that, as with his sister, he chose a time when his wife was fast asleep for most of his sexual activities with her.

His wife could satisfy his two main wishes—either to be a child or a woman—since she appeared to him as a phallic mother. Because of his fear of being abandoned by his parents and left to the desolation of utter loneliness, he developed a set of fantasies in which he identified with both his father and mother. Through the psychotic condensation of fantasying himself as a hermaphrodite, he maintained a state of pseudo independence of dangerous parental images. This simultaneous playing of both roles created unbearable tension, for which marriage seemed to be an emergency solution. In the marital interaction it was possible to project his own discordant roles onto his wife.

It need hardly be mentioned to what extent this marriage satisfied the patient's masochistic craving for rejection and frustration. Every day and night spent with his young wife brought a repetition of his fantasied rejections of the past. Even after most of these mechanisms had become clear to him and he had started divorce proceedings, he still tried time and again to approach his wife. His explanation was characteristic: "When I was a child," he said, "I was convinced my parents wanted to leave me. I could not allow that to happen since I did not recognize the existence of anyone else. I wanted to force them to stay with me and tried to maintain my ties with them through hatred."

Months after his divorce the patient disclosed recurring amorous fantasies centered around his wife. His comments about it on one occasion were symptomatic of the depth of his ego regression: "I am like in a state of trance. I just noticed it. The world does not exist, so I can maintain my fantasy about loving my wife. I am a child and she will take care of me. I am in a bubble, in a globe."

As to the last point of our inquiry, which is concerned with the influence of therapy on the course of such marriages, it is theoretically possible that psychoanalytic therapy with both schizophrenic partners may restore the marriage to a healthy relationship. Yet our experience shows that such an outcome is the exception rather than the rule. Rarely, are both husband and wife sufficiently motivated to undertake deep, reconstructive psychotherapy. Since the marital choice is based on pathological motives, it can hardly be expected that with the reconstruction of the ego they will maintain their original choice.

I have seen situations where psychoanalysis helped one of the partners to free himself from the pathological relationship, so that eventually he was able to enter into a more adequate marriage. This happened in the case of Mrs. S. and also in that of Mr. X. It also holds true for patients who belong to the preschizophrenic group themselves and who marry as a way of acting out and satisfying their deeply regressive dependency needs. With the progress of therapy and liquidation of those infantile needs, the acting out loses its meaning and urgency.

Summary

The following points emerge from our analysis of the patterns of marriage between schizophrenics:

1. Their object choice is often determined by the wish to act out a number of unconscious fantasies, not infrequently of a contradictory nature.

2. These fantasies and, accordingly, the object chosen are supposed simultaneously to fulfill certain isolated infantile wishes as well as offering a safeguard against them. Thus we may say that both in the choice of objects and in coping with ambivalent feelings and dangerous aspects of infantile love relationships, the schizophrenic displays mechanisms utilized by the infantile ego—repetition and denial.

3. Owing to the fragmentation of his ego, the schizophrenic, in choosing his mate and establishing his pattern of marriage, can pursue contradictory aims. The persistence of archaic forms of ego organization compels him to seek in his mate a reincarnation of infantile aspects of early love objects.

4. Since only a part of his ego remains attuned to reality, the schizophrenic attempts to pursue in his marriage goals completely at variance with reality.

5. In keeping with his general tendency toward restitution of the loss of reality, the schizophrenic will use his marriage for this purpose. However, the realization of this restitutive function of the marriage is jeopardized by the deficiency of his reality testing and, in cases of a psychosis of his partner as well, by the inadequacy of the latter and his independent pursuit of his own unrealistic and archaic goals.

The Alcoholic Spouse

RUTH FOX, M.D.*

EXCESSIVE DRINKING of alcohol by one or both partners in a marriage ranks high as a contributing cause of broken homes. Divorce, financial insecurity, unemployment, illness and preventable accidents, desertion and maltreatment of children, juvenile delinquency, prostitution, and minor and major crimes are among the consequences.

One need only visit the Home Term Court in New York City to get an idea of the amount of suffering and anguish caused by this distressing phenomenon of our modern life. It is estimated that forty per cent of the problems brought before this court are directly or indirectly attributable to alcoholic excesses.

The fact that there have been so many varying definitions of alcoholism indicates the general confusion regarding this disorder. However, certain characteristics do stand out as distinguishing the compulsive drinker from the normal or even exces-

* President, New York City Medical Society on Alcoholism; Vice President, National Committee on Alcoholism.

sive social drinker. Seliger defines an alcoholic as "one whose drinking definitely interferes with one or more of his important life activities—in business, in the home, in the family and in the community." [19] Wolberg stresses the "need of the alcoholic to drink in order to function, or his consideration of it as his chief source of pleasure." [24] Diethelm defines it as "that condition in which an individual harms himself or his family through the use of alcohol and either cannot be made to realize it, or, realizing it, no longer has the will or strength to overcome his habit." [8] Seldon Bacon, the sociologist, describes alcoholism as "a condition appearing in adult individuals which may be labeled a disease." [2-4] Among its characteristics he lists the following: the developing and eventually chronic compulsion to drink alcohol as a prop for meeting ordinary life problems; a series of responses, such as blackouts, sneaking drinks, specific guilt reactions, solitary drinking, "benders," tremors, and the like; increasing social maladjustment in all aspects of life—job, relations with parents, marriage, clubs and other activities.

My own working definition is quite simple. If a patient is *unable* to stop drinking after two or three drinks, he is almost certainly an alcoholic. While the alcoholic can abstain totally for varying periods of time, he is unable to drink moderately. For him it is all or nothing and one drink is usually the beginning of a spree. In other words, drinking for the alcoholic is compulsive in nature, once he has started. He is driven by unconscious forces he does not understand and against which rational judgment and will power are helpless.

Called our Public Health Problem Number Four, alcoholism ranks just behind heart disease, cancer and venereal disease. Its victims in the United States number approximately four million. According to Bacon, six and a half per cent of the total adult male population is alcoholic. It would seem that men are more prone to become alcoholics than women, since the ratio in our country is about one woman to six men. However, with the alarming increase of alcoholism in women in the past two dec-

ades, the ratio may be changing. That this is a cultural phenomenon and not an inborn difference in susceptibility can be
shown in the wide variations in this ratio in different parts of
the world. In England where women drink without condemnation in the pubs along with the men, the ratio is 1:2. In Nor-
way where there is not this permissiveness toward women's
drinking, it is 1:23. In Switzerland it is 1:12.

The greatest cost to the nation, as Yahraes points out is in the
"breakup of families, the dulling of fine minds, and the warping
of lovable personalities." [25] To me, most appalling of all is
the price children and mates must pay in bewilderment, humiliation and often physical neglect and abuse. Indeed, for every
alcoholic there are probably four or five other persons who suffer
from his or her excessive drinking: wives, husbands, mothers,
father, children, employers, employees and friends.

It cannot be denied that powerful social forces are continually being exerted upon all of us to drink and to do so frequently.
Our culture fosters and encourages drinking, not of course
with the intention of making alcoholics, but that is often the end
result. There is also a conflicting attitude toward drinking—one
that regards it as evil—and this has been pointed out by Myerson
who refers to it as the "social ambivalence toward alcohol." [15]
When regarded as sinful, however, it has the added appeal of
the forbidden and can be put to the service of rebellion.

Yet some people are more susceptible to these social pressures than others. Wherein lies this vulnerability to the on-
slaught of alcohol? Efforts to prove that a constitutional or
hereditary defect lies at the root of alcoholism have been generally unfruitful. To be sure, after prolonged heavy drinking
there are marked physiological disturbances which require
medical attention during therapy. But these should be looked
on as the result rather than the cause of the alcoholic state.
The vulnerability would seem to be psychological in origin, consisting of a defect in personality structure.

Generally speaking, there are three types of alcoholics: the secondary addict, the primary addict and the situational addict. In those alcoholics who have been called secondary addicts— i.e. those in whom the condition appears after many years of heavy but controlled drinking—this defect may be slight and pass unnoticed for quite a while, the individual concerned making an apparently good adjustment in educational, occupation, social and marital spheres of his life.

People who eventually become secondary addicts do not seem to have a special psychological need for alcohol early in life. They may drink heavily on occasion, but not more so than many in their group who do not later become alcoholic. Many excel in their work, and their drive, originality and creativeness may be above average. As the years go by, the amount of liquor they consume may gradually increase, until they find they have developed a dependence on it. Their drinking having become compulsive, the amount they drink and the appropriateness of time and place are no longer matters of conscious will; they have lost control. Alcohol is resorted to more and more frequently in times of stress.

The personality defect becomes more and more apparent, causing increasing difficulties in interpersonal relations. And to this are added the very real problems engendered by the alcoholism itself: the loss of earning power, the loss of prestige and the regard of friends, and often the loss of affection and love from those closest to them. They become self-centered, egocentric, suspicious and under-socialized. Many can be helped to return to their previous level of accomplishment when they give up alcohol. However, because of their relatively high tolerance for alcohol, enjoyed in early life, and the almost daily drinking of large quantities over many years, an appreciable number may show mental and physical changes in the form of delirium tremens, Korsakoff's psychosis, polyneuropathy, advanced liver disease or a deterioration of the personality.

Bacon describes the primary addicts as persons who have been psychoneurotic throughout their lives, with alcoholism starting at an early age, often in their teens.[2-4] These individuals were obviously maladjusted on an emotional level prior to their compulsive drinking. They might have been introverted or insecure with respect to interpersonal relations, or excessively dependent. At first, alcohol may have provided great relief. It undoubtedly helped diminish their shyness, allowed them to be aggressive, or give full play to self-pity. The prognosis of treatment for the primary addicts, whose drinking is symptomatic, is not as good as for the secondary. Because of greater personality defect, they have never reached as mature a level of adjustment, and in most cases they have not been able to express themselves adequately in socially acceptable ways.

Bacon describes a third type of alcoholism which he calls "situational." This may occur in a normal adult who is placed under some unusual emotional strain, such as the loss of a loved one, a severe accident, war, financial reverses, or loss of prestige. Though the drinking in these people may be indistinguishable from the other forms of alcoholic drinking, with help these persons can pull out of their pathological drinking. Usually, they are unable to revert to social drinking, though there may be cases in this group where this is possible.

Methods of treatment and prognoses vary according to the type of drinker involved. The alcoholics with the best prognosis are certainly the situational drinkers. When a solution is found to their basic problem, the pathological drinking may be discontinued. The prognosis for secondary addicts is also quite good, since they were reasonably well-adjusted in their earlier adult lives. When they have given up alcohol, they can resume the more mature pattern of living.

The primary addicts, who have become addicted in very early life, are basically extremely passive and dependent people. Their problem is really one of growing up, and this may take a long time. The social misfit or "skid row" variety of drinker rarely wants to stop, and literally has no life without alcohol. The prognosis of psychotics who drink abnormally will depend upon the resolution of their psychoses. It is important to note that none of these people who have reached the stage of pathological drinking—i.e. those who are true alcoholics—can ever again drink in a controlled fashion. The aim of therapy is to teach these people how to lead a satisfactory life without the aid of alcohol.

The question has often been asked as to whether there may not be a special type of personality that becomes alcoholic, and many psychological tests have been performed in an attempt to find an answer. Of the 850 patients to whom I administered Antabuse in the last six years, about one third were given a battery of psychological tests by Mrs. Leatrice Styrt Schacht.* The findings revealed that many different types of personalities are capable of becoming addicted. This is not surprising since alcohol may be used to "solve" the most diverse kinds of emotional conflicts.

However, the test subjects did show many characteristics in common. Among these were an extremely low frustration-tolerance, inability to endure anxiety or tension, feelings of isolation, devaluated self-esteem, a tendency to act impulsively, a repetitive "acting out" of conflicts, often an extreme narcissism and exhibitionism, a tendency toward masochistic self-punitive behavior, sometimes somatic preoccupation and hypochondriasis and often extreme mood swings. In addition, there is usually marked hostility and rebellion (conscious or unconscious) and repressed grandiose ambitions with little ability to persevere. Most show strong (oral) dependent needs, frustration of which will lead to depression, hostility and rage.

It should be noted that the degree of maturity in an alcoholic varies just as it does in the rest of us. As in other neurotic conditions, the more traumatic the childhood, the more serious will be the personality defect underlying the alcoholism, and consequently the more intensive the therapy required.

---

* Psychologist, Alcoholics Treatment Center, New York City.

Treatment of the symptom of overindulgence of alcohol is not enough. The patient must acquire a more mature attitude so that he will not find it necessary to resort to alcohol.

Before discussing the incompatibility of a good marriage and alcoholism in one or both of the marital partners, it would be well to define which segment of the population of alcoholics we are describing. The three groups of men on whom we have valid statistical studies relative to marital status are arrested inebriates, homeless or "skid row" derelict itinerants, and those presenting themselves for treatment in community out-patient clinics for alcoholics.

There is wide variation in the degree of general social integration in these three groups, as well as in their marital status.[21] Also, the probability of successful outcome of treatment varies according to the group to which a patient belongs. Arrested inebriates are often psychopaths who are usually not amenable to our present methods of treatment. In contrast, Straus and Bacon, in an analysis of over 2000 male patients who sought help in nine out-patient alcoholism clinics in various parts of the United States, describe a different type of alcoholic.[21] This group shows a "relatively high degree of social and occupational integration . . . over half were married and living with their wives. . . . The percentage who had never married was no greater than normal expectancy . . . More than eighty per cent were under fifty years of age; a fourth were under thirty-five. An impressive number had sought treatment before reaching a state of personal and social disintegration. . . ."

For purposes of this study, I shall discuss the marital problems of this third group of uncontrolled drinkers. In spite of the fact that marriage has been entered into by this group as frequently as in the normal population, four times as many of their marriages have been dissolved as would be expected in the normal population.

Probably no marriage with an alcoholic can be considered a happy one. There may be periods of relative harmony, but there is such a basic inadequacy in the one who drinks (and surprisingly enough often in his partner too) and lack of faith in human beings that the mutual trust and sharing necessary for a good relationship are absent. The chief aim of a compulsive drinker, once he has started, is to continue drinking—that of the spouse to prevent it. These diametrically opposed attitudes inevitably lead to quarrels, recriminations and/or psychological withdrawal on the part of one or both of the partners.

It is extremely unlikely that an alcoholic, once he is caught up in the egocentricity which is an inevitable by-product of his illness, can actually love another person in a mature sense. To him love is rather a desire to be loved and cared for, the primitive kind of loving an infant feels for the mother.[6] In the alcoholic arrogance or the drive to be omnipotent is at bottom a desperate attempt to deny and cope with his underlying feeling of dependency, helplessness, fear and impotence. The alcoholic's desire to force someone to accede to his wishes and love him in spite of what he does makes him an extremely difficult person to live with.

In order to stifle feelings of inadequacy, hostility, isolation and helplessness—an aggregate of emotions Horney calls the basic anxiety[11]—the alcoholic, from time to time, tries out various, often conflicting maneuvers, each designed to raise his self-esteem. On one occasion he will be ingratiating, charming, even fawning, on another, hostile, grudging, even cruel. At a later date he may withdraw into himself, becoming aloof, cold and seeming to need no one. This unpredictability is another reason the alcoholic is hard to live with.

Another way used by the alcoholic to overcome his basic feeling of helplessness, as Portnoy has shown, is the building of an unconscious fantasy that he *is* whatever he *wishes* to be.[17] Since he wants to be a grandiose creature, he deludes himself into thinking he is this glorious person. Then, regarding himself as unique and special, he thinks he is entitled to preferential treatment. The arrogant pride, the inordinate claims for special and unlimited privileges, the feeling of being entitled to unconditional happiness and love, the conviction that he should be free of responsibility for his actions—all these elements operate unconsciously, for the most part, when the individual is not drinking.

Thus the alcoholic feels constantly angry, hurt, put upon and abused. His demeanor may be one of smoldering discontent. Sometimes he shows his anger and tremendous resentment, but more often he does not dare do so because his need for people is so great and his fear of desertion or retaliation so devastating. He often has a burning desire for vengeance on a world that he feels has treated him shabbily. Those nearest him—the ones he needs most—are usually the chief targets for his venom. Hence wives, mothers, sisters, fathers, employers come in for most of the abuse.

It is not surprising that from forty to sixty per cent of all alcoholics come from the disturbed background of an alcoholic family. The children of alcoholics tend to be neurotic because the sense of security so necessary for the building of a strong and independent ego is rarely found in their household. The child can be utterly bewildered by the sudden shift in behavior of the alcoholic. A parent who is often affectionate, understanding and fun-loving when sober may become morose, demanding, unreasonable, touchy, noisy and even cruel and violent when drunk. Or a naturally reserved and somewhat withdrawn parent may become sloppily sentimental and seductive in the early stages of a drinking bout, or hilariously and embarrassingly exuberant. He or she may spend money wildly and make extravagant promises which are impossible to fulfill. The frequent swing from high hopes to shattering disappointments may build up in the child such a basic distrust that all his later intimate relationships will be distorted.

When the father is alcoholic, a son will find it difficult to attain any stable identification, so that an ambivalence in feelings may beset him throughout life. He may react by cowering in the corner, outwardly conforming, or he may rebel and become defiant or delinquent. He may become overly attached to the mother, but deny this out of fear. Or he may turn against both parents because of their incessant quarreling. He may even come to feel that the non-drinking mother is to blame for the father's drinking, and thereby identify masculine independence with drunkenness.

The daughter of a drinking father will sometimes hate him and side with her mother. But she may also continue to love him desperately and feel the mother is to blame. Each time he drinks, there is a feeling of deep personal rejection. "If he *really* loved me," is her thought, "he wouldn't drink." She may also confuse independence with drunkenness, and this may be responsible for her own alcoholism in later life.

Difficult as it is to have an alcoholic father, it is often possible for the mother to shield the children somewhat from the full impact of the situation. With an alcoholic mother, this is rarely if ever the case. Because of the closeness of children to the mother, they are apt to suffer irreparable damage. The seemingly sudden withdrawal of love produced by liquor in any quantity may lead to deep and lasting feelings of rejection, isolation and abandonment. Though they may become actively defiant, they often react with deep feelings of guilt later on for their defiance.

By the time children have become conscious of persons outside their immediate family, they have also become aware of the condemnation of the alcoholic by society, and they react with deep shame and humiliation. They feel different, estranged and isolated. When their alcoholic parent is jeered at, they may try to defend him out of love. Usually, however, they cannot bring themselves to do so, and they feel deep guilt at what seems to them a betrayal of the parent.

Because our culture condemns alcoholism in women more than in men, there are many hidden, lonely drinkers among women. Among working and professional women, as well as those in active social circles, women may drink quite openly with their male friends. Yet in their early years, in spite of their apparent freedom, they tend to be ashamed of making a spectacle of themselves. Somehow they are able to curtail the amount they drink until they can get home and unreservedly finish off a bottle. Housewives may be able to hide their drinking from their husbands for quite some time. Not being subject to the discipline of fixed hours at an outside job, they may drink small amounts all day long. For a long time such a woman can continue with her household duties, her shopping and the care of her children, all the while being just a little befuddled. Her drinking is usually a closely guarded secret, with the money for the alcohol taken from the household budget.

The emotional withdrawal from husband, children and friends may be so gradual that the husband does not suspect a thing. Although somewhat uneasy at the way the marriage seems to be slipping away, he may accept his wife's explanation that she is overworked or that the children are too much for her. It may be an enormous shock to him to come home one day and find her drunk, with the work undone and the children afraid or running wild. Because of the double standard, the husband is usually disgusted and angry.

A good deal of therapy may be necessary for him, as well as for his wife, before he will accept her condition as an illness. I have found men generally less patient and accepting of alcoholism and less willing to learn about it than are the wives of alcoholic men. They are more apt to pack up and leave an alcoholic wife whom they feel they can no longer love. Sometimes the children too are abandoned along with the wife, but battles in the courts over custody of the children are frequent. The wives of alcoholic men, on the other hand, will make almost any sacrifice to help their husbands once they have learned to look upon them as sick individuals. This is partly because of their greater tendency to mother and sympathize with the husband, sensing that he cannot help himself, and is also due to the greater permissiveness in our culture toward drinking among males. Also, there is the fact that the wife is usually financially dependent on the husband for her own and the childrens' support.

A number of interesting personality studies have been made of the wives of alcoholics. [7, 10, 18, 23] Baker points out that though the wife is not responsible for the alcoholism of her husband, she may be one of the reasons for his continued drinking in spite of treatment.[5] Since her personality disturbance may be even more serious than his, she is equally in need of psychotherapy or counseling. Futterman shows that some women seem to need to be married to weak, dependent, alcoholic males.[10] One woman stated that she would not divorce her third alcoholic husband because she knew very well that she would only marry a fourth alcoholic.

Such a woman is often the daughter of an alcoholic father and a dominating mother. Her ego ideal, through identification with the mother, is that of the powerful woman. She sees herself as indispensable, as capable of playing the role of both mother and father to her children, and tends to push the drinking husband further and further out of the family. Though this is her conscious picture of herself, unconsciously she feels inadequate to live up to her ego ideal as either wife or mother. It is only by feeling superior to her husband and keeping him inferior to her that she can deny her own basic inadequacy. Frequently, as the alcoholic becomes abstinent, this type of woman no longer has her foil and may decompensate emotionally herself, reacting with severe depression or other neurotic disturbance.

Boggs also stresses the need to treat the "alcoholic marriage," since each of the partners seems to be striving to make the other play the role which would meet his individual needs.[7] According to this author, good adjustment in the wife of an alcoholic is rare indeed. "The uncanny ability of the alcoholic to seek in marriage an equally immature and needful person" is emphasized.

Pointing out that the wife often tries to keep her husband inadequate to justify what seems to be a lack of love for her on his part, Price describes the wives of alcoholics as basically dependent people, nervous and hostile, despite an appearance of adequacy and capability.[18] These women marry hoping to find strong, supportive, dependable persons on whom to lean. On discovering how incapable the alcoholic is of filling this role, they react to the apparent rejection with hostility and resentment. Unable to consider the spouse as a person with needs and wishes separate from her own, the wife tends to put more and more demands on her husband, making him feel and actually become less and less adequate.

Whelan, in a perceptive and lovely way, goes into considerably more detail in describing the various types of wives of alcoholics found in a family service agency.[23] What the alcoholic contributes to a disturbed family situation may be all too obvious in terms of noise, angry neighbors, delinquent children, irate landlords and visits from the police. All this turmoil may give the impression that the wife is the innocent victim. Actually, the alcoholism may be a red herring for the neighbors; the neurotic wife may bear equal responsibility for the unhappy marriage.

Among the husbands of alcoholic women, there are the following types: the long-suffering martyr who mothers and spoils his child-wife, the husband who leaves furiously but comes running back, the unforgiving and self-righteous husband, and the punishing, sadistic variety. There is also the dependent male who expected to find another mother in his wife and who is hurt and bewildered at finding that the woman he married— the one who put up such a show of self-confidence—has become just as dependent as he through her alcoholism. Then, of course, there is the "normal" man who wakes up in dismay to find himself with an alcoholic wife.

Sometimes we find alcoholism in both partners of a marriage. No deeper tragedy could befall their children. All is chaotic and unpredictable, with frequent violent and bitter quarrels. The pattern varies greatly in these marriages. When one is sober, the other may drink out of retaliation or relief when the partner sobers up. Both partners may start drinking together socially and end up in bitter quarreling. Not infrequently a spouse becomes alcoholic in an attempt to keep his alcoholic partner company.

If both are alcoholic, one may recover and find himself or herself just as intolerant of the other's drinking as those who had never been alcoholic. Many persons in Alcoholics Anonymous have partners who are still drinking but refuse to recognize it as a problem. In some cases they are alcoholic; in others, merely heavy social drinkers. In either case the alcoholic finds it difficult to continue abstaining.

Like the alcoholic, the family of an alcoholic also goes through various stages in trying to adjust to his abberant behavior. Jackson describes three stages in cases where the husband is the alcoholic.[12] Early in the marriage there may be an occasional overstepping of bounds with heavy drinking. This may be embarrassing but is not considered too serious, and both partners tend to minimize its importance. As the frequency of such occurrences increases, the wife begins to feel humiliated and ashamed. She curtails their social life, hoping in that way to diminish her husband's temptation to drink. As a result, invitations are received less often and the ensuing social isolation leads to too-close family interactions. The wife is under the impression that she has somehow failed in her marriage, and both partners begin to feel deep resentment.

What Mrs. Mann calls the "home remedies" are persistently tried, despite their evident failure to control drinking.[14] Liquor is locked up and bottles hidden. Money is withheld and charge accounts cancelled. The family tries moving from the city to the country, or vice versa. At first the couple discuss the situation with "sweet reasonableness," later with anger and recrimination. Emotional appeals—"How can you do this to me?" "Where is your self respect?" "Think of the children"—are as ineffective as everything else. This is because the alcoholic, though sincere in his promises, is *unable* to stop without outside, expert help.

During this stage all is chaos. The children become involved and are bewildered. There is hostility, frustration, fighting, threats of leaving. The wife reacts to the alcoholic's violence by cringing in terror, retaliating, or calling the police. There is economic anxiety, for often in this phase the alcoholic works only intermittently.[18] The wife may fear for her sanity and the emotional effect on the children. The alcoholic himself begins to think he is "insane," since he cannot understand why he does the things he does.

Becoming too ill to recover at home, the alcoholic begins the trek to hospitals or sanatoria or doctors. He is frequently arrested for disorderly conduct or chronic drunkenness. Finally, when he has "hit bottom," he may try Alcoholics Anonymous. The wife may go with him to the A.A. meetings or to the meetings of the A.A. Family Groups.[1] She begins to learn techniques of management and gets much support from identifying herself with other wives who have had the same problem.

If the alcoholic husband does not recover through A.A., psychiatry, Antabuse or the Church, the wife may find courage enough to take some constructive long-range steps. She may precipitate the crisis necessary to "bring her husband to his senses" and force him into treatment by separation or divorce. The husband may be very difficult at this time—physically violent, threatening to kidnap the children, turning up drunk at the wife's place of employment, withholding all support if the wife is not working, or pinning her down with threats of suicide.

It requires great fortitude and courage for the wife to carry through her plan, but it is often necessary to do so for her husband's recovery. This need to interrupt the dependency pattern of certain male alcoholics by withdrawal of emotional, physical and financial support is well described by Myerson.[16] The woman who is being leaned on may be the mother, sister or wife, sometimes a friend. In any case, shielding the patient against the consequences of his drinking may seriously impede his recovery.

If the husband does sober up and the family does take him back, there may be unsuspected basic problems still to solve. Having been in charge for so long, the wife may dislike relinquishing her power. Then, too, the children are not used to including their father in their plans and may resent his sudden assumption of authority over them. They may hesitate to go to him for advice, and, if they do, the mother may feel left out and resentful.

In his new sobriety, the husband may expect endless praise from his wife for his "brave comeback." Yet she may continue to berate and belittle him for his past actions, being unable to wholly forgive him. If he spends too much time at A.A. meetings, she may grow angry and again feel rejected and abused. However, with the help of psychotherapy or A.A. for both husband and wife, it is remarkable how many families do manage to attain a large measure of happiness. While sobriety is a prerequisite, it is not the total goal. The whole family may need help in reorienting itself to a new way of life. In this regard, the wife must bear in mind that in our culture the man should be at least an equal if not the master. He cannot function successfully without a sense of personal dignity.

As to the general principles of therapy,[9] the medical man, especially the family physician, is probably one of the first persons to be consulted, and his attitude toward the patient as well as the family may be of the utmost importance in eventual recovery. The idea that an alcoholic is necessarily a difficult hospital patient has been disproved. When accorded the courtesy and respect due any patient, an alcoholic is grateful and almost always cooperative. This has been the experience in a number of hospitals that have been sponsored by A.A., such as Knickerbocker and St. John's in New York, and also in some of the private hospitals that accept alcoholics. Today, with proper medical management, patients in the acute stages of drinking or in the hangover stage can be made relatively comfortable and quiet. Hence they are far less difficult and demanding than alcoholics in the past.

Generally, the best results are obtained by using a combination of several therapies at once. This is what is known as the "total push." The modern treatment of the acute alcoholic episode has been greatly simplified. The various nutritional deficiencies of the patient must be met as adequately as possible. This means that he needs food, especially protein, salt, fluids, vitamins and some sedation.

Treating the emotional problems underlying alcoholism means a long and complicated process of rehabilitation, using the interdisciplinary approach. An analyst alone cannot accomplish it. We need to draw on the help of psychologists, social workers, nurses, experts in vocational guidance and recreation, hospitals and rest homes, schools, colleges, courts, churches, social and fraternal organizations, Alcoholics Anonymous—in fact, all that our complicated society has to offer. Because of the need for this teamwork, the individual alcoholic is best treated in special out-patient centers or clinics, such as those now operated in a few of the states.

The two most important adjuncts to psychotherapy of the alcoholic are Alcoholics Anonymous and the use of Antabuse. The remarkable accomplishments of Alcoholics Anonymous attest to the value of this social and religious approach. Often the greatest service to the alcoholic and his family is getting the patient to accept the fellowship of this group. Not every alcoholic can or will accept A.A. But for those who can, there are so many immediate as well as lasting advantages that every effort should be made to see that a suitable contact is made. Except in the completely irreligious and in the deeply neurotic or psychotic alcoholic, A.A. can either be the mainstay of treatment, or the only treatment needed.

Psychiatrists, ministers and sociologists have studied this group approach of A.A. and have tried to interpret its success. The movement has drawn a great deal from both medicine and religion and is a powerful therapeutic weapon. As one of the founders has said, "A.A. is a society where men and women understand each other, where the clamors of self are lost in our great common objective, where we can learn enough of patience, tolerance, honesty, humility and service to subdue our former masters—insecurity, resentment and unsatisfied dreams of power." [22]

Tiebout believes that through the force of religion in an atmosphere of hope and encouragement, the typically egocentric, narcissistic alcoholic personality, dominated by defiant individuality and drives for omnipotence, undergoes a profound change by the acceptance of the A.A. program. The negative characteristics of aggression, hostility and isolation are replaced by peace and calm and a lessening of inner tension.

This change or conversion may come suddenly with cataclysmic strength, but in most cases it is effected gradually through the living out of the program. The defiant individual no longer resists help, but accepts it and allows such positive feelings as love, friendliness and contentment to replace the former feelings of restlessness and irritability. It is claimed that fifty to seventy per cent of the alcoholics who go to A.A. remain abstinent. Perhaps those who are able to accept the program of A.A., without at the same time receiving psychotherapy, are less neurotic than many patients seen in analysis.

The least sick can do well with a combined Antabuse and

Alcoholics Anonymous approach. The more neurotic need either counseling and retraining or group psychotherapy. The most neurotic need analytic therapy in addition to Antabuse.

The psychoanalysis of alcoholics can be greatly aided by the use of Antabuse. The drug should be instituted in those cases with other neurotic manifestations in addition to alcoholism. Without Antabuse, alcoholics frequently cannot tolerate the frustrations of analysis and are apt to revert to drinking during difficult periods of the treatment. Early in the analysis one must tackle the drinking problem itself. Alcohol can be looked upon as an outside force against which the patient carries on an endless struggle. Though thrown to his knees again and again, he will not and often he cannot surrender. This may be partly because he will not relinquish the pleasure of the infantile regression which alcohol allows. But it is also because he feels it a defeat he will not tolerate.

As previously noted, the alcoholic has grown up with the mistaken idea that it is "manly, smart and sophisticated" to drink and only a weakling cannot "hold his liquor like a man." One of the most important tasks of psychotherapy is to bring the patient to the point of accepting his inability to drink. This is only a part of his total unwillingness to accept reality and it may take months or years of painstaking analysis before the partipotent attitude,[20] in contrast to the omnipotent, of the normal adult becomes fixed in the personality.

The degree of ability to accept reality is in direct proportion to the ego strength of the individual. Since the ego strength is low in an alcoholic, we must be prepared to give him much support in the early months of treatment. This may be attained through a strong transference relationship, through the ego building of A.A. or through the protective technique of Antabuse.

Just as there is no single explanation for alcoholism, so there is no single treatment. The results of various types of therapy carried on during the last ten years have shown remarkably similar rates of recovery, regardless of the method used. The three points of similarity in the various successful techniques seem to be a sincere desire on the part of the patient to get well, an acceptance of the alcoholic as being ill rather than perverse, and a hopeful attitude toward his ability to arrest the condition.

References

1. *Alcoholics Anonymous Family Groups: A Guide for the Families of Problem Drinkers.* New York, P.O. Box 1475, Grand Central Annex 17, 1955.

2. Bacon, S. D., "Mobilization of Community Resources for the Attack on Alcoholism." *Q.J. Studies on Alcohol,* VIII, 473-97, December, 1947.

3. ———, "Alcoholism: Its Extent, Therapy, and Prevention." *Federal Probation,* XI, No. 2, 1947.

4. ———, "Alcoholism: Nature of the Problem." *Federal Probation,* II, No. 1, 1947.

5. Baker, S. M., "Social case work with inebriates." In "Alcohol, Science and Society." New Haven, *Quarterly Journal of Studies on Alcohol,* Lecture 27, 1945.

6. Balint, Michael, "On Love and Hate." *International Journal of Psychoanalysis,* Vol. 33, Part IV, 1, 1952.

7. Boggs, M., "Role of Social Work in the Treatment of Inebriates." *Quarterly Journal of Studies on Alcohol,* IV, 557, March, 1944.

8. Diethelm, O., *Treatment in Psychiatry.* New York, The Macmillan Company, 1936.

9. Fox, R., "Psychotherapeutics of Alcoholism." *Specialized Techniques in Psychotherapy.* J. L. Despert and G. Bychowski, eds. New York, Basic Books Inc., 1952.

10. Futterman, S., "Personality Trends in Wives of Alcoholics." *Journal of Psychiatric Social Work,* XXIII, 37-41, 1953.

11. Horney, K., *Neurosis and Human Growth.* New York, W. W. Norton & Co., 1950.

12. Jackson, J., "The Adjustment of the Family to the Crisis of Alcoholism." *Quarterly Journal of Studies on Alcohol,* XV, 562-86, December, 1954.

13. Landis, B. Y., "Some Economic Aspects of Alcohol Problems." *Q. J. Studies on Alcohol*, VI, 59-91, 1945; IX, 259, 1948.

14. Mann, M., *Primer on Alcoholism*. New York, Rinehart & Co., Inc., 1950.

15. Myerson, A., "Roads to Alcoholism." *Survey Graphic*, February, 1945.

16. Myerson, D. J., "An Active Therapeutic Method of Interrupting the Dependency Relationship of Certain Male Alcoholics." *Quarterly Journal of Studies on Alcohol*, XIV, 419, September, 1953.

17. Portnoy, I., "Psychology of Alcoholism." From a lecture given in 1947 for the Auxiliary Council of the Association for the Advancement of Psychoanalysis.

18. Price, G. M., "A Study of the Wives of 20 Alcoholics." *Quarterly Journal of Studies on Alcohol*, V, 620-27, 1945.

19. Seliger, R. V., "The Problem of the Alcoholic in the Community." *American Journal of Psychiatry*, XCV, 701, 713, November, 1938.

20. Silverberg, W. V., "Factors of Omnipotence in Neurosis." *Psychiatry*, XII, 387-98.

21. Straus, R. and Bacon, S. D., *Alcoholism and Social Stability*. New Haven, Hillhouse Press, 1951.

22. Tiebout, H. M., "Therapeutic Mechanisms of Alcoholics Anonymous." *The American Journal of Psychiatry*, C, 468-73, January, 1944.

23. Whelen, T., "Wives of Alcoholics: Four Types Observed in a Family Service Agency." *Quarterly Journal of Studies on Alcohol*, XIV, 632-41, 1953.

24. Wolberg, L., *Medical Hypnosis*, Vol. I, p. 328. New York, Grune & Stratton, 1948.

25. Yahraes, H., "Alcoholism Is a Sickness." Public Affairs Pamphlet, No. 118, 1946.

General References

Rado, S., "The Psychoanalysis of Pharmacothymic Drug Addicts." *Psychoanalytic Quarterly*, II, 1-23, 1933.

Research Council on Problems of Alcohol, *The Scientific Approach to the Problem of Chronic Alcoholism*. New York, 1946.

State of Vermont, Alcoholic Rehabilitation Commission, "As a Family You Should Know These Facts About Alcoholism." Pamphlet 174, 1955.

The Measurement of Psychological Factors in Marital Maladjustment

MOLLY HARROWER, Ph.D. *

ALTHOUGH PSYCHODIAGNOSTIC EXAMINATIONS to evaluate the intellectual and personality characteristics of a patient are a well established procedure as an adjunct to psychotherapy, little has been reported on such personality appraisals of two or more persons involved in some interpersonal difficulty and tested at approximately the same time. Several years ago, a study was reported by me in which four members of a family, the parents and two sisters, were all tested with a view to throwing light on the problem of the elder girl, the "patient." [1] The interrelatedness of the problems of the two sisters became apparent against the background of the parents' personalities. More recent work

* Director, Psychological Testing Program, University of Texas Medical Branch, Galveston, Texas.

about to be reported by Peck, Harrower and Margolin will show the interrelatedness of the personality characteristics in members of "gangs" of juvenile delinquents who have appeared in the Children's Court of Manhattan as first offenders. These studies have thrown light not only on each individual child, but on the role within the gang which his personality prescribes that he should play.

Findings in both these investigations impressed me with the inadequacy, in cases involving close interpersonal relations, of psychological examinations on one individual alone. Joint examinations would seem to be indicated, particularly in those cases of marital difficulties where, for some reason, one member of the pair has become the "patient" while the other remains outside the therapeutic and evaluative situation. Studies on both partners in such circumstances might give helpful information.

This report deals with eighty persons, forty couples, who had been referred for psychodiagnostic examination by a psychiatrist or physician because their presenting problems involved marital difficulties. These forty couples received a full psychodiagnostic work-up, and in addition to the personality appraisal of each individual separately, a comparison of findings was made in terms of the respective depth or seriousness of the disturbance as reflected in the test findings. Sometimes suggestions as to the type of difficulties which might be expected from two persons with these respective test profiles were formulated. Occasionally, prognostic expectations were hazarded. These, however, pending experimental evidence, had to be very tentative.

This investigation was in progress over a period of eight years. During these eight years, the tests which were administered were augmented. This was inevitable since new and valuable tests become available ever so often and the psychodiagnostician is, or should be, ever alert to the possibilities of obtaining more relevant material from his examinations.

Methods

Most of the forty couples received the following six tests during the examination: The Verbal Wechsler-Bellevue Intelligence Test, the Szondi test, the Rorschach test, the Figure-drawing test, the Most Unpleasant Concept test, the Holsopple-Miale Sentence Completion Test. All the couples received at least four tests, so that every personality appraisal, and every comparison between the members of a given couple, was the result of material derived from various test instruments. In every case the same tests were given to both partners. Thus material from couples tested some eight years ago, while based on four rather than six tests, was always uniform.

The material from the referring psychiatrist, relating to the couples' background, difficulties, psychiatric status and final recommendations for therapy, ran the whole gamut from complete case histories through letters highlighting the essentials, to brief comments, to no information at all. It is clear that material such as this, which accumulates and grows without initially being part of planned research, must be subject to the hazards of the inevitable variations in individual procedures. Regardless of the extent to which the research psychologist may wish to correlate findings in each case and at every stage with a psychiatric opinion, the fact remains that in clinical practice this is not always possible.

When the final assessing of these eighty cases began, a short questionnaire was sent to each of the physicians responsible for referring these patients. The questionnaire dealt with the following items: *1.* Are these persons still married? *2.* Did one partner or both go into therapy? *3.* If so, for how long? *4.* Did this therapy materially alter, modify or have some influence on the marital situation?

Replies from the majority of the referring physicians accounted for thirty-four of the forty couples. The results were as follows: still married, eighteen; status not known, ten; divorced, five; one partner deceased. The figures in regard to the effectiveness of therapy are as follows: therapy materially altered status of marriage, fourteen; therapy modified status of marriage, eight; no therapy given or no effect, seven; died, one; not estimated, four. For two couples who were counseled by me, a detailed follow-up over a period of several years was available; for ten others sufficient information was obtained so that the test findings could be verified.

Results

The psychological material was originally assessed in the usual manner, and then was re-assessed in the light of the existing marital maladjustments. A general over-all appraisal from all the six tests resulted in a statement about each couple which fell into one of the following eight categories:

1. Both partners appear psychotic.
2. The husband appears psychotic, or borderline, the wife less disturbed.
3. The wife appears psychotic, or borderline, the husband less disturbed.
4. The husband appears more seriously disturbed, but not psychotic.
5. The wife appears more seriously disturbed, but not psychotic.
6. Both appear equally, and seriously disturbed, but not psychotic.
7. Both appear equally somewhat disturbed.
8. Both appear to fall within normal limits for the total population.

This classification was made twice. In some cases, there was an interval of as much as eight years between the original and the second assessment. This is important in that it constitutes a check or validation on the procedures and on the examiner. The second assessment was based on the raw material or the raw scores, and again a judgment in terms of the respective seriousness of the partners' difficulties was given. It is interesting to find that in only one of the eighty cases was the second judgment different from the first. In the second assessment in this case both partners were considered to be psychotic whereas previously only one had been listed as psychotic, the other as "seriously disturbed."

These findings may be classified as follows: Both partners psychotic, three; man psychotic or borderline, woman less disturbed, ten; woman psychotic or borderline, man less disturbed, one; man more disturbed, no psychosis, twelve; woman more disturbed, no psychosis, five; both seriously disturbed, five; both somewhat disturbed, two; both within normal limits, two.

The relatively large incidence of cases where the husband is either psychotic (ten cases) or the more seriously disturbed (twelve cases) is somewhat startling. These two groups together account for twenty-two husbands, or over half of the total examined. In contrast, those cases where both persons are equally disturbed, although there are four categories from which to draw, amount to only twelve couples; and those cases where the wife is more seriously disturbed, total only six cases.

Associated with the finding that in the majority of cases the husband is more disturbed, is another item of interest. On several cases, several years apart, the examiner's notes carry a comment to the effect that despite the fact that the wife appeared to be the more seriously disturbed during the interview, her test findings seemed to belie this. After consulting several of the therapists who had referred such couples, there appeared to be evidence which would support the hypothesis that the *least disturbed* partner comes to therapy first. Pressure is often exerted on the "patient" to enter therapy by the partner, who, it is later discovered is the more disturbed mate.

One therapist, for instance, wrote: "In the cases which I referred to you certainly the healthier person went into therapy

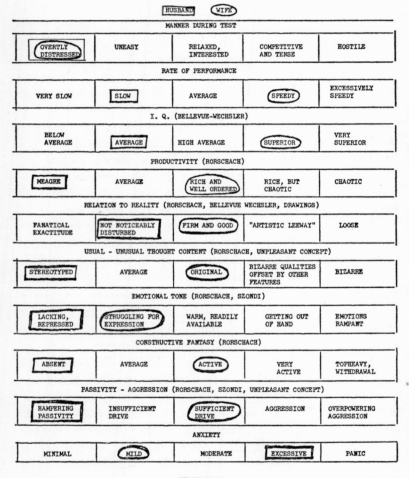

[HUSBAND] (WIFE)				
MANNER DURING TEST				
[OVERLY DISTRESSED]	UNEASY	RELAXED, INTERESTED	COMPETITIVE AND TENSE	HOSTILE
RATE OF PERFORMANCE				
VERY SLOW	[SLOW]	AVERAGE	(SPEEDY)	EXCESSIVELY SPEEDY
I. Q. (BELLEVUE-WECHSLER)				
BELOW AVERAGE	[AVERAGE]	HIGH AVERAGE	(SUPERIOR)	VERY SUPERIOR
PRODUCTIVITY (RORSCHACH)				
[MEAGRE]	AVERAGE	(RICH AND WELL ORDERED)	RICH, BUT CHAOTIC	CHAOTIC
RELATION TO REALITY (RORSCHACH, BELLEVUE WECHSLER, DRAWINGS)				
FANATICAL EXACTITUDE	[NOT NOTICEABLY DISTURBED]	(FIRM AND GOOD)	"ARTISTIC LEEWAY"	LOOSE
USUAL - UNUSUAL THOUGHT CONTENT (RORSCHACH, UNPLEASANT CONCEPT)				
[STEREOTYPED]	AVERAGE	(ORIGINAL)	BIZARRE QUALITIES OFFSET BY OTHER FEATURES	BIZARRE
EMOTIONAL TONE (RORSCHACH, SZONDI)				
[LACKING, REPRESSED]	(STRUGGLING FOR EXPRESSION)	WARM, READILY AVAILABLE	GETTING OUT OF HAND	EMOTIONS RAMPANT
CONSTRUCTIVE FANTASY (RORSCHACH)				
[ABSENT]	AVERAGE	(ACTIVE)	VERY ACTIVE	TOPHEAVY, WITHDRAWAL
PASSIVITY - AGGRESSION (RORSCHACH, SZONDI, UNPLEASANT CONCEPT)				
[HAMPERING PASSIVITY]	INSUFFICIENT DRIVE	(SUFFICIENT DRIVE)	AGGRESSION	OVERPOWERING AGGRESSION
ANXIETY				
MINIMAL	(MILD)	MODERATE	[EXCESSIVE]	PANIC

FIGURE 1

Sample chart giving summary of text findings

first . . . and in another case which I have not referred but have recently seen the so-called 'crazy' wife has needed only short-term therapy while the husband is much more seriously disturbed."

Because these cases were examined at irregular intervals, and because there was no awareness on the examiner's part at the time of the "group as a whole," this trend was not apparent. However, during the initial interviews, it was clear that persons who appear equally disturbed or confused over their problems in the clinical setting made a very different impression when their test findings were assessed. It has not been unusual to find that the more distressed person at the testing interview turns out to be less disturbed in the tests themselves.

In order to focus attention on this, and to make possible a quick contrast of the performance of psychologically related persons, a summary sheet was developed which gives some of the highlights from the various tests.[2] One of these sheets (Figure 1) shows how these over-all appraisals can be compared at a glance. It also demonstrates that identical behavior in the testing situation does not mean comparable test performance. In this instance, one partner is shown on the test as an individual with superior potentialities, the other is shown as more seriously disturbed and inhibited.

This over-all appraisal from the total test findings is, however, only a small part of what can be gleaned from this accumulation of test material. Each of the six tests used yields details which have been fully evaluated. In this report only the details from the Verbal Wechsler-Bellevue are presented.

This well-known intelligence test is composed of several subtests revealing specific abilities or lack of them, such as the individual's fund of informational knowledge, his comprehension, reasoning and judgment, his memory for digits, his arithmetical ability, his capacity for abstractions or generalized thinking. Since the performance scales were not used, the "verbal intelligence quotient" (IQ) results from totalling the scores on all these subtests. But it is very important to remember, since it is so frequently forgotten, that one and the same I.Q. can be obtained from very different subtest scores.

Figure 2 illustrates the method of scoring the Wechsler-Bellevue test, and may serve as an orientation for the non-psychologically trained reader to the "scattergram." Both "A" and "B" have I.Q.'s of 101. Nothing could be more "average" or "ordinary" than such an I.Q. of 101 which is just on the midline and represents average intelligence. Yet these two individuals could not have more varied psychological equipment or

A (I.Q. 101)

Equivalent Weighted Score	Information	Comprehension	Digit Span	Arithmetic	Similarities
18	25	20		14	23-24
17	24	19	17	13	21-22
16	23	18	16	12	20
15	21-22	17		11	19
14	20	16	15		17-18
13	18-19	15	14	10	16
12	17	14		9	15
11	15-16	12-13	13		13-14
10	13-14	11	12	8	12
9	12	10	11	7	11
8	10-11	9			9-10
7	9	8	10	6	8
6	7-8	7	9	5	7
5	6	5-6			5-6
4	4-5	4	8	4	4
3	2-3	3	7	3	3
2	1	2	6		1-2
1	0	1		2	0
0		0	5	1	

B (I.Q. 101)

Equivalent Weighted Score	Information	Comprehension	Digit Span	Arithmetic	Similarities
18	25	20		14	23-24
17	24	19	[17]	13	21-22
16	23	18	16	12	20
15	21-22	17		11	19
14	20	16	[15]	10	17-18
13	18-19	15	14		16
12	17	14		9	15
11	15-16	12-13	13		13-14
10	[13-14]	11	12	8	12
9	12	10	11	7	11
8	10-1	9			9-10
7	9	8	10	6	8
6	7-8	7	9	5	7
5	6	5-6			[5-6]
4	4-5	4	8	4	4
3	2-3	[3]	7	3	3
2	1	2	6		1-2
1	0	1		2	0
0		0	5	1	

I.Q. 101   A   I.Q. 101   B

Weighted score 48 at 21 years.   Weighted score 48 at 21 years.

FIGURE 2

have been more different in their behavior. "A," for instance, is a uniformly-functioning, stable, mediocre individual; "B," is a psychotic girl. Although both these scores add up to 101, they add up to it by virtue of very different sub-totals, representing very different abilities. The apparent equality of their "intelligence quotient" (IQ) is arbitrary, and may even be misleading.

Let us imagine that these two people, "A" and "B," are husband and wife, and visualize these two graphs superimposed. With their respective psychological equipment, certain flagrant difficulties would be bound to exist. "B," who is now the hypothetical wife, has extremely low scores on the test for comprehension and the test for abstract thinking. She would not understand what would be relevant action, nor would she be able to think logically. On the other hand, she has an excellent rote memory and is smarter at figures than "A." Although "A" and "B" both have the same score on the information test, it would be found, if the actual records were studied, that "A" loses points because he cannot define some of the more difficult or technical words (questions at the harder end of the scale), while "B" loses points because of failure to know the number of weeks in a year or the date of Washington's birthday, but excels in the more erudite information at the harder end of the scale. While our conception of these two people as a pair is hypothetical, they are actual persons who have been examined. Comparison of their Wechsler-Bellevue "scattergrams" illustrates a method by which some psychological factors in marital partners may be assessed.

As a result of such comparisons in the present study, six specific "patterns of discordancy" emerged; six variations, that is, between the "scattergrams" of the two partners. Moreover, it was possible to place each of the forty couples under the general heading of one or another of these six patterns. In some cases the specificity of these patterns allowed for a kind of check-up within the therapeutic situation, directly related to the kind of difficulty reflected in the tests. For instance, in one case the wife, who was the patient, repeatedly spoke of difficulties which related to her lack of ability in some areas, her "dumbness" in contrast to her husband's exceptional abilities in these same areas. Not until the husband himself was tested did the full implications of these difficulties, both in their realistic and neurotic aspects, become clear.

Figure 3 illustrates one "pattern of discordancy." It was obtained from the husband and wife just mentioned. In this, and in all other scattergrams, the black dots relate to the husband's scores while the outlined figures refer to the wife's. This particular "scattergram" illustrates several points. First we notice the husband's unusually high, uniform performance. Here is an individual whose score is only a few points below the maximum

Equivalent Weighted Score	Information	Comprehension	Digit Span	Arithmetic	Similarities
18	25	20			
17	24	19		13	21-22
16	23		16	12	20
15		17		11	19
14	20	16	15		17-18
13	18-19	15	14	10	6
12	17	14		9	15
11	15-16	12-13	13		3-14
10	13-14	11	12	8	12
9	12	10	11	7	11
8	10-11	9			9-10
7	9	8	10	6	8
6	7-8	7	9	5	7
5	6	5-6			5-6
4	4-5	4	8	4	4
3	2-3	3	7	3	3
2	1	2	6		1-2
1	0	1		2	0
0		0	5	1	

FIGURE 3

Equivalent Weighted Score	Information	Comprehension	Digit Span	Arithmetic	Similarities
18	25	20		14	23-24
17	24	19	17	13	21-22
16	23	18		12	
15				11	19
14	20	16	15		17-18
13	18-19	15	14		16
12	17	14			15
11	15-16	12-13	13		13-14
10	13-14	11	12	8	12
9	12	10	11	7	11
8	10-11	9			9-10
7	9	8	10	6	8
6	7-8	7	9	5	7
5	6	5-6			5-6
4	4-5	4	8	4	4
3	2-3	3	7	3	3
2	1	2	6		1-2
1	0	1		2	0
0		0	5	1	

FIGURE 4

obtainable. On three of the subtests he does achieve the maximum scores. Both within our series and even in the population at large, such a uniform high level performance is very unusual. This individual (the only one in the series) was not a patient and offered to be tested in the hope of throwing light on his wife's difficulties after she had been in therapy for some time. In contrast to this, the wife's score is both much lower in general level (she would fall into the Average, rather than the Very Superior Group of the total population), and it also appears that her performance is much more erratic. While she does well on the abstract thinking, she does extremely poorly in the arithmetic.

It might be well at this point to explore a little more fully what high or low scores in the respective subtests may mean. Although we cannot go into this in any great detail, it has been reliably demonstrated that extremely low scores on "arithmetic" are almost invariably due to anxiety which sometimes reaches panic-like proportions in this subtest, rendering the individual completely incapable of functioning. A sharp drop in the "arithmetic" scores, therefore, is almost always an index that the individual may be at the mercy of disrupting and disorganizing anxiety. Particularly, as in this case, when an individual is capable of logical thinking (note the much higher score on the "similarities") one may assume that one can contrast the capacity of the individual when not anxious with performance under the stress of panic.

While a drop in the score in "arithmetic" is, let us say, a deviation from a better level and is, therefore, a "hopeful" sign, a sharp drop in the "similarities" usually indicates a more serious type of disturbance in which anxiety does not play a prominent role. For, when the score in the "similarities" is very low, it indicates that the individual is showing a distortion in thinking, that there is a disturbance in his relation to reality. Metaphorically, then, one might say the framework within which he envisages things has itself become warped and distorted.

A lower score in "information" may mean that the ordinary things in life no longer are of much interest, because the individual has withdrawn into a world of his own. This is shown by failure to know or register easy facts, as opposed to lack of educational opportunities or lack of intellectual curiosity. The latter would leave its mark by the lack of knowledge of more specific items of learning or culture.

Low scores on "comprehension" usually reflect, in somewhat the same way as a lower score in "abstract thinking," that the individual is unable to appraise or understand what the pertinent thing to do is in situations of various kinds. This again reflects a more serious disturbance.

A low score on "memory" may also reflect the impact of anxiety, since the individual is too frightened to listen carefully enough to the numbers he should repeat, or it may reflect a lack of concentration due to the pressure of internal problems so that the mind is never "blank" enough to register and hold onto the incoming information.

Returning now to the husband and wife, whose "scatter-grams" are shown in Figure 3, it is clear that to some extent their difficulties relate to their different levels of intelligence, but their difficulties also relate to the steadiness and the intellectual uniformity of the husband and the erraticness on the part of the wife. More specifically, however, we may say (since this was very clearly brought out over a period of time in the therapeutic sessions) that the wife becomes panic-stricken. She accentuates her own "dumbness" in any situation where it appears to her that she is being shown up at a disadvantage, as when she feels incompetent or not as well informed as her husband. The discrepancy between their scores on the test for "information" are significant in this respect. Therapeutically, the hopeful aspects of this situation are the relative closeness in their scores on logical thinking ("similarities") and the understanding of situations ("comprehension"). In actual fact, the outcome here was a very favorable one, and after several years of therapy this wife's "scattergram" looked much less erratic and her I.Q. rose to a point where she reached the High Average to Superior Group of the total population.

For purposes of discussion this type of joint "scattergram" will be referred to as *Panic Introduced by Feelings of Incompetence*. It is clear that it also represents different levels of functioning and extremes of "scatter" and "lack of scatter."

Figure 4 represents a husband and wife who have an identical type of scatter, who really have parallel scores. However, one occurs at a lower level of intelligence than the other. In this instance, a follow-up was possible, which showed that the more intelligent and less disturbed husband went into brief psychotherapy which was considered to be unusually effective by the

Equivalent Weighted Score	Information	Comprehension	Digit Span	Arithmetic	Similarities
18	25			14	23-24
17	24	19	17	13	
16	23	18		12	20
15	21-22	17			19
14	20	16	15		17-18
13	18-19	15	14	10	16
12	17	14		9	15
11	15-16	12-13	13		13-14
10	13-14	11	12	8	12
9	12	10	11	7	11
8	10-11	9			9-10
7	9	8	10	6	8
6	7-8	7	9	5	7
5	6	5-6			5-6
4	4-5	4	8	4	4
3	2-3	3	7	3	3
2	1	2	6		1-2
1	0	1		2	0
0		0	5	1	

FIGURE 5

Equivalent Weighted Score	Information	Comprehension	Digit Span	Arithmetic	Similarities
18	25	20			23-24
17	24	19	17	13	21-22
16	23	18		12	20
15		17		11	19
14	20	16	5		17-18
13	18-19	15	14	10	16
12	17	14		9	15
11	15-16	12-13	13		3-14
10	13-14		12	8	12
9	12		11	7	11
8	10-11	9			9-10
7	9	8	10	6	8
6	7-8	7	9	5	7
5	6	5-6			5-6
4	4-5	4	8	4	
3	2-3	3	7	3	3
2	1	2	6		1-2
1	0	1		2	
0		0	5	1	

FIGURE 6

therapist. This couple is still married, and, according to the therapist, in this case therapy "materially altered the existing marital conditions." Incidentally, the other members of his group also showed a satisfactory outcome, subject to therapy in the case of the more intelligent partner. We believe this distribution, *Parallel Scores at Different Levels,* to be a "benign" pattern, subject to successful amelioration with therapeutic intervention.

Figure 5 shows the type of interrelated "scattergram" which we have called *Cut-Throat Competition.* It will be noted that both graphs are at the extreme upper limits and that each varies only from the hypothetical ideal score by exceedingly small amounts. The basic competitiveness of these two partners presented rather fundamental difficulties which can be understood from the scores. Apparently, too great a similarity in intellectual achievements does not allow sufficient "room" for individuality to develop. Both these partners vied for the same role in the relationship. In terms of Gestalt psychology, neither partner seemed able to become the "ground" against which the "figure" could show up. Both partners needed to be the "figure" at all times. It is an interesting sidelight that both individuals confessed rather shamefacedly after the examination that they had attempted to find out what tests were to be used in order that they might memorize some of the answers.

Figure 6 reflects a difficulty, perhaps the most significant of all, which involves gross distortions in thinking on the part of one partner and the panic-like reaction of the other. In this case, the husband who brought the wife to therapy was not examined until much later, but when he was examined an underlying schizophrenic process (reflected here in completely atypical or distorted thinking) was brought to light. It then became evident that the wife's acute anxiety (note the low score on "arithmetic") related directly to this realistically frightening, but insidious, quality in the husband's thinking. From this and other cases, we would make the suggestion that this becomes a vicious cycle in which the panic-stricken partner becomes more and more disorganized and unable to cope with (because he is explicitly unaware of) the distortions in thinking of the other. It should be noted that there is a tremendous discrepancy in the "scattergram" between the scores for abstract thinking, and a wide discrepancy, but in the opposite direction, in "arithmetic." *Conceptual Distortions Producing Panic* might be suggested to describe this type of paired "scattergrams."

Figure 7 represents similar types of distortions in both partners. It will be noticed that both individuals here show atypical thinking (low scores on "similarities") and that both are outstandingly accurate in "arithmetic," and that the rest of the graphs differ very slightly, one from the other. Interestingly enough, this does not produce panic reactions and unlike the

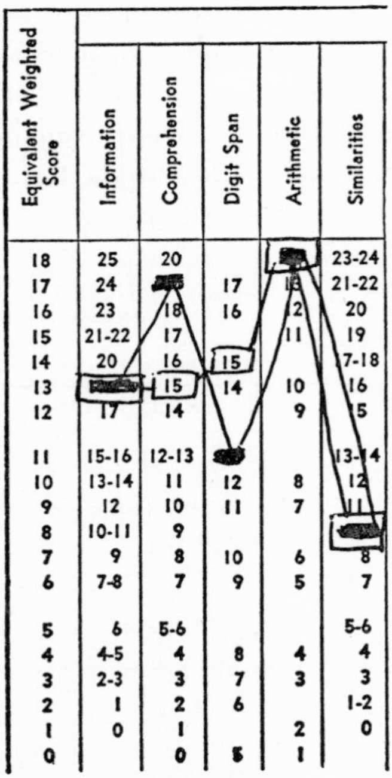

FIGURE 7 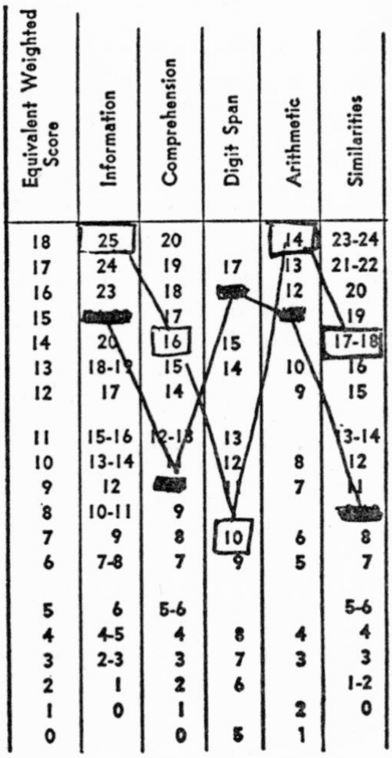   FIGURE 8

graphs in Figure 6, the marital adjustment of such couples appears much more amenable to therapeutic help. One couple in this group was followed up for a five year period following brief therapy for each with the same therapist. *Similar Distortions of Thinking* might be said to describe these patterns.

Finally, Figure 8 reflects the chaotic distribution and wide scatter of two gifted but borderline partners. It will be noted that there is extreme erratic performance in both cases, with a variation of 10 points for one partner and 8 for the other. There is considerable discrepancy between the scores for virtually every one of the five subtests. Such couples pose very difficult therapy problems and such graphs might be described by the term *Excessive and Deviant Scatter*.

In Table 1 I have summarized a classification of marital partners disclosed by the psychological testing of forty pairs of husbands and wives. It is significant that the psychological testing methods used were sufficiently quantitative to make their classification possible. Furthermore, the case histories, where available, indicate that these tests are fairly reliable as indices against which the results of psychotherapy may be evaluated.

TABLE I—*Types of psychological patterns observed in "scattergrams" of 40 pairs of maladjusted husbands and wives as examined by the Wechsler-Bellevue test*

TYPE	NO. OF PAIRS	SUGGESTED PROGNOSIS
Panic induced by feelings of incompetence	7	Possible improvement with therapy
Parallel scores at different levels	9	Good with therapy
Cut-throat competition	5	Doubtful
Conceptual distortions producing panic	11	Doubtful
Similar distortions in thinking	4	Good with therapy
Excessive and deviant scatter	4	Doubtful

Discussion

For centuries, physicians, clergymen, lawyers and thoughtful laymen have been aware of the probability that psychological factors are involved in marital discord. Vague and intuitive analyses of such factors, with lack of quantitative methods of measuring them, have been serious obstacles in the rational management of marital maladjustment. Modern methods of personality appraisal, using standardized and quantitative testing methods, may furnish the badly needed foundation for the systematic development of a rational and effective way of handling marriage difficulties due to psychological factors involved in the personalities of marital partners.

It will be necessary to carefully test and follow many more pairs for at least five years in order to determine the extent to which the patterns so far noted may have prognostic value in the marriage situation. The finding in this series of forty pairs that more than half of the husbands were psychotic or seriously disturbed appears to be significant statistically. Of psychological significance is the observation that the more disturbed partner is usually the one who comes for treatment last.

Of what avail, one may well ask, are such statistics if they cannot be translated into some kind of immediate help for puzzled people? How can these findings, as well as the results of other research studies, best be utilized for the practical help of individuals facing marital problems?

Psychologists may rightly be criticized if their interest in these problems remains exclusively within the province of their ivory towers. But psychologists, along with other scientists, are often unwilling to make suggestions based on relatively scanty findings for fear of violating standards of accuracy. On the other hand, the patients who have cooperated in such test programs are justified in their impatience with the slow methods of research.

They often ask questions which force the psychologist to tackle the problem of how to apply and extend his new methods of diagnosis so that they may be of value to as many people as possible.

The following questions are typical: Would premarital psychological tests (like premarital Wassermans) do anything to prevent a neurotic choice of mate or neurotic interaction? On the basis of the psychological tests, are there people who should not consider marrying? Does the material the psychologist collects at the time of acute interpersonal conflict point to ways in which marital difficulties can be resolved? Is everyone more or less neurotic, and, if so, what hopes for the institution of marriage can we develop through educational measures which may contribute to better adjusted couples?

As to whether premarital psychological tests can help in the prevention of neurotic choice in marriage, it would seem that young couples contemplating marriage are turning to the psychologist with just such hopes in mind. The usual reason for seeking psychological help and consultation at this point is that one or the other of the pair has become increasingly apprehensive about the forthcoming marital step. They look to the psychological tests to supply the answer as to whether or not their contemplated move is a wise one. In a score of such consultations given over a period of years, it has become clear that the test answers should never amount to a categorical yes or no in response to the individual's queries. Rather, what can be seen in strong relief and communicated to the given couple are the kinds of psychological problems that are bound to arise in the normal course of events when they link their lives within the framework of marriage.

It has been found helpful and advantageous, for instance, to discuss the personality characteristics of each partner, first with each one alone and then in a combined session. In this way the individuals involved may become alerted to possible pitfalls where they may expect their basically different types of reaction to introduce conflict. Merely taking the tests does not produce an immunity to conflicting situations. But being alerted to the possible areas of trouble has been found to be very advantageous.

Take, for example, the case of Joan and George whose tests showed some marked differences in psychological equipment. For George, accuracy and a punctilious concern with detail was of great importance. Joan, on the other hand, approached all situations in a much more general way. Basic issues concerned her, not their detailed elaboration. She would respond quickly and spontaneously, perhaps intuitively, on the basis of evidence that George would find completely insufficient or unsatisfactory.

In the course of counseling sessions it turned out that each resented, criticized and yet envied these qualities in the other. While George inwardly criticized Joan and would soon have made that criticism overt and explicit, he nonetheless envied her for the swiftness and sureness of her convictions and the ease with which she tackled a problem, in contrast to his own painstaking approach. The essence of the counseling lay in the fact that each was shown the other's test scores, revealing the pressures each was under. The advantages and disadvantages of both orientations or attitudes were disclosed, and each became aware of the danger signals in themselves when they exaggerated those traits which in moderation were not pathological.

Dr. Grey Walter, in his book *The Living Brain*, has described ways in which the actual functioning of the brain may be shown to differ in two individuals, with correspondingly different ways of reacting to and assessing everyday situations.[3] "Fortunately," states Dr. Walter, "extreme types are rare, but when two people display unreasonable and irreconcilable differences of approach to a question, before concluding that this is due to innate antagonism or incompatability of purpose, a discrepancy in their ways of thinking may be worth while looking into." According to Dr. Walter, "ways of thinking" are the direct counterpart of the different "brain waves" his instruments show. The psychological tests, through a different medium and at a different level, also reflect these different ways of thinking. To this extent, premarital testing and psychological counseling may actually forestall potential psychological disturbances.

The second question is: On the basis of psychological tests, are there people who should not consider marrying? Perhaps the answer is that an interim period might well be spent by one or both of the prospective partners in clearing up certain psychological difficulties that are likely to trigger off marital conflicts.

The following example is a composite one with features taken from several actual cases. The girl suggested that her fiancé take psychological tests because she felt that he was always ready to criticize her, although he was very fond of her. In the course of testing both prospective partners it transpired that the girl was projecting, or throwing outward onto someone else, some of the fears that really related to herself but which she was loathe to face. It further turned out that part of her difficulties had to do with her fear of an overly severe and somewhat emotionally distant father. Although she did not recognize it, she was seeing in her fiancé many of her father's psychological characteristics, as envisaged by her when she was a small child.

The psychological evaluation pointed to the fact that the girl's chances for a happy married life with anyone would depend largely upon her becoming aware of what some of the old anxieties and fears of her father were doing to her, and, equally important, of her unwillingness to accept them as being her problems. The answer, then, is not that these two "should not marry," but rather that they should not marry until the girl has had a chance to work out her problems with a therapist.

Findings based on any of these psychological tests should not, however, be considered all conclusive. Couples seeking advice must not be led to believe that all difficulties will be eliminated in their subsequent married life if they follow the psychologist's recommendations. Nor does the recommendation to postpone a marriage mean that the marriage will not be successful when ultimately consummated. In what is by far the greatest number of cases, it would seem that those persons who have enough insight to know that it will help to understand themselves better are more likely to work toward a constructive solution of their marital problems.

The next question is: Does the material which the psychologist collects at the time of acute interpersonal conflict point to ways in which marital difficulties can be resolved? Regarding the material produced by an individual in the tests, and the constructive solutions of marital difficulties, it would seem that there is a very direct relationship between this material and the kind of help that can be given. In the short space of time in which the individual responds to psychological tests, he provides the experienced examiner (who may subsequently assume the role of psychological counselor) with an immense amount of material about himself. Often the individual tested is quite unaware of the symbolic meaning of much that he reveals through the medium of the tests. Yet very frequently these answers go to the heart of the psychological difficulties the person is facing.

For example, through the medium of the Ink Blot pictures it may become clear to the psychologist that a man's unconscious ideas and fantasies about the woman he plans to marry are completely at variance with what he believes himself to feel and think about her. The flights of his imagination as seen in the stories he tells and the figures he finds in the Ink Blots may be about quarrelsome, ugly or threatening women. Yet he will consciously deny that he ever feels this way toward women in general, let alone his wife to be.

Through these responses, it is possible to make people realize that they are reflecting some of their hidden attitudes, and to show them the extent to which their feelings are confused or ambivalent. This method, known as projective counseling, has proved of great value in working toward a solution of many types of marital problems.

Finally, the question is asked: If everyone is more or less

neurotic, what hope is there for the institution of marriage, or can steps be taken to bring about better adjustment by educational changes? The kind of answer the psychologist would most probably give to this question can be seen from the foregoing discussion. To be forewarned is to be forearmed. Awareness of personality characteristics, awareness of one's blind spots, awareness of what things are apt to send one into a tailspin—all these balance the individual's perspective so that he is less likely to be caught up in confusing and anxiety-producing situations which he cannot control.

A clearer understanding of what an individual wants out of life, particularly out of his or her married life, is an important ingredient in choosing a partner. Greater self-knowledge before marriage and throughout marriage can only be helpful in insuring the individual's lasting happiness.

Summary

Six "patterns" formed by the combined scattergrams of husband and wife in maladjusted couples have been shown to exist. The distribution within our series was as follows: *1. Panic introduced by feelings of incompetence,* eighteen per cent; *2. Parallel scores at different levels,* twenty-three per cent; *3. Cutthroat competition,* fourteen per cent; *4. Conceptual distortions producing panic,* twenty-seven per cent; *5. Similar distortions in thinking,* nine per cent; and *6. Excessive and deviant scatter,* nine per cent.

At least one example from each group has been followed carefully over a period of time so as to correlate the problems which appeared in the therapeutic interviews with the type of difficulty which would be expected from the intellectual pattern as shown on the Wechsler-Bellevue test. These patterns would appear to have value in a quick assessment of the type of difficulties which the couples are facing, and may be used within the marriage counseling situation to give objectivity to the problems presented. The combined patterns have also been used as a measuring rod to demonstrate changes which have occurred after therapy. More cases are needed with full follow-up in order to estimate the prognostic significance of the classes of patterns disclosed.

References

1. Harrower, M. R., "Neurotic Depression in a Child." *Case Histories in Clinical and Abnormal Psychology*. A. Burton and R. E. Harris, eds. New York, Harper & Brothers, 1947.

2. ———, *Appraising Personality*. New York, W. W. Norton & Co., Inc. 1952.

3. Walter, Grey, *The Living Brain*. New York, W. W. Norton & Co., Inc. 1953.

Research on Human Movement Response in the Rorschach Examinations of Marital Partners

———•◆•———

ZYGMUNT A. PIOTROWSKI, Ph.D.*

STEPHANIE Z. DUDEK, M.A.†

RESEARCH IN THE AREA of marital adjustment is complicated by the fact that a multiplicity of factors which are difficult to isolate and analyze enter into the contraction and maintenance of a marriage. Moreover, aside from the reasons given by the marital partner for his dissatisfaction with his marriage, we have few if any objective criteria for measuring marital happiness.

Ideally, fundamental compatibility is the ability of two people to understand and communicate with each other in the same degree and on an equal level of complexity, and to accept each

* Research Psychologist, New Jersey Department of Institutions and Agencies; Adjunct Professor of Psychology, New York University.
† Psychologist, Department of Psychiatry, Vanderbilt Clinic, New York City.

other at the level of the least changeable traits of the personality. Fundamentally compatible people accept those aspects of each other's personality that seem to be basic, fixed and least liable to change, or adjust to each other's needs or the demands of society. Fundamental compatibility can exist on a healthy or a neurotic base, and is generally not conscious.

Since, considering the multiplicity of factors involved, it is clearly impossible to define happiness in marriage, our basic question is focused on an understanding of why couples who show every sign of unhappiness—conflicts, quarreling, sexual frustration, open dissatisfaction with each other as mates and with the marital state—continue to maintain their marriages instead of seeking divorce or legal separation? What are the ties which hold such a marriage together? The usual answers include the ingredient of love, children, financial considerations, religious taboos, etc.

Although important and valid, these factors are probably only secondary. If it is true that people love their desires more than the objects of their desires, the primary reasons for maintaining a difficult marriage must be of a very personal and basic nature. It can be assumed that the couples who stay together under such circumstances do so because of certain imbedded personality traits, or because the partners have a fundamental compatibility on a deep unconscious level.

The basic factors that hold people together in a marriage fraught with discord are likely to be neurotic ones, and the compatibility is apt to be on a neurotic level. Such a marriage remains permanent because the partners consciously or unconsciously fulfill each other's neurotic needs—those needs on which the equilibrium of their personalities and their social adjustment depends. In fact, this equilibrium seems to be achieved or maintained only by the unconscious adjustment the partners make to each other's neurotic needs. In a genuinely happy or well adjusted marriage, the needs are healthy. But whether healthy or neurotic, this unconscious adjustment to each other on a deep level makes for the fundamental compatibility so essential in marriage.

The aim of any personality is to seek and maintain a state of psychological equilibrium and freedom from debilitating tension —this state depending on personality needs and adaptation to the world. A successful personality adjustment is one which has a high degree of freedom from anxiety and neurotic conflicts which tie up energy and result in somatic breakdowns or states of turmoil, fear and insecurity.

The neurotic defenses of any person, regardless of how self-defeating they may be in the long run, are his efforts to maintain psychological balance, preserve his integrity as a person and protect himself from real or imagined annihilation. Even the reasonably healthy person must develop defenses against being physically and emotionally hurt, against the coercions and pressures of the world.

The methods used to maintain health and integrity determine character structure. A very neurotic person's defenses against danger will be warped by his distortion of reality, by his irrational fears and projections and peculiar attitudes. A healthy person's defenses are rational and appropriate to the challenge. Since no one is sufficient unto himself, the neurotic defenses are essentially against the threat imposed by other people on one's psycho-biological energy economy. Thus the defenses become interwoven with the people with whom the individual has established close and intimate relationships.

Of these, marriage is generally the closest and deepest relationship, maintained over a very long period of time if not for life. Even the choice of a partner will depend to some extent on one's neurotic needs and defenses. Presumably a marriage will be maintained so long as it does not vitally threaten the basic character structure, or so long as it is effective in supporting it in a state of reasonable comfort.

One of the mechanisms through which a psychological balance is maintained is complementation. It implies that two people are attracted to each other on the basis of their healthy or neurotic needs for the purpose of maintaining and developing their habitual needs and goals. Some character structures maintain their equilibrium better through identification with a similar character structure, in which case similarity will be the basis of choice; others get along better when mated with a character structure different from their own. There are undoubtedly still other mechanisms of mutual character adaptation, none existing in a strictly pure form.

The type of relatively superficial compatibility that is assessed by the questionnaire method or a paper-and-pencil prediction scale does not concern us here. Such procedures have supplied us with very little relevant data in evaluating marital adjustment. The very nature of fundamental compatibility compels us to use techniques which are able to reflect the deeper motives and drives within the personality. For this reason we have chosen the Rorschach method as the experimental procedure for measuring the degree of similarity or difference in basic psychological traits,[3] limiting ourselves essentially to one component of the method—the human movement response, symbolized by the capital letter M.

Rorschach considered the M his most important component, and his concept of its meaning is his most original contribution to the psychology of personality measurement. The discussion of the significance of the various possible numerical values of the ratio of M to CR (color responses) occupies more space in Rorschach's writings than any other single component of his method.

We concentrated on an analysis of the M for several reasons. In the first place, our subject groups differed much more in the quantity and quality of their M than in any other component. Secondly, the M is said to reflect the cultural influences absorbed by the individual and determine more than any other Rorschach component what makes the individual a distinct personality. Rorschach emphatically made the point that persons differing in their number of M are divergent personalities, and that psychological communication between them on a deep level is difficult and only partial. Finally, a comparison of the entire Rorschach records of the subjects would have been a formidable and complicated task, and even then no clear patterns differentiating between the groups could be convincingly demonstrated.

Wide experience, including repeated examinations of neurotic patients undergoing intensive and successful psychoanalytically-oriented psychotherapy, indicates that the type or quality of M changes very slowly and gradually, and that it is most resistant to change.[2] Assuming that the personality traits revealed by the M are similarly resistant to change, and are therefore deep-seated and powerful motivating forces, the study of the M for the purpose of explaining the duration of close psychological relationships, such as those which develop in marriage, is desirable as well as justified.

Rorschach and his followers looked upon the M as an indicator of repressed tendences which are not expressed in overt behavior. In 1937, however, this view was challenged and a new definition of the M was advanced. It implied that, on the contrary, the psychological tendencies revealed by the M press for direct and overt motor manifestations.[1] This new interpretation does not exclude the possibility that the environmental and subjective forces which caused the M tendencies to originate may be unconscious. Even the M tendency itself may be unconscious in the sense that the individual may be unaware of it and may have repressed his desire and ability to become aware of it. Yet this does not prevent the M tendency itself from pressing for outward manifestation.

According to Rorschach, the M is positively correlated with imagination, with a liking for inner living, with interest in the world of ideas and cultural values, with intensive rather than extensive relationships with people. The larger the number of

M, absolutely and by comparison with the color responses, the greater the preference for very close contacts with few people and dislike for easy and rather superficial contacts with many people.

Persons with many M are inclined to develop a definite style of life. They cultivate self respect, they have a tendency to think first and act later in the face of trouble, they tend to be secretive regarding their motives, they prefer to rely on themselves rather than on others in difficult situations, and they like to base their security on the development and growth of their personal assets, knowledge, professional skills, intelligence, self-control, and the like.

Another important implication of the M is that it discloses the fundamental attitude, the prototypal role in life which the individual assumes toward others when personally vital matters are involved. Self-reliance and compliance are the two most important fundamental attitudes, and there are many combinations of these two.

Self-reliance is not synonymous with activity and initiative, nor is compliance synonymous with passivity and inactivity.[1] Self-reliance, as disclosed by extensor M, goes together with self-confidence and a tendency to be a leader. The chief need of a compliant person, as revealed in the flexor M, is to have a benevolent protector who assumes responsibility for whatever is done. This delegation of responsibility to someone felt to be benevolent and psychologically stronger than oneself enables many a compliant person to become active and display initiative.

To reiterate, the M reflects a person's way of life, his concept of himself in relation to the world, the concept that dominates his adaptation to others in personally vital matters. Although one does not always behave according to the traits revealed by the M, people who live together intimately cannot differ greatly in the number and quality of their M and expect to experience harmony. There are too many vital and important decisions of a partnership nature in marriage which require living in accordance with one's deep character traits as revealed by the M.

This presupposes a similar way of life for the married couple, an attitude toward the world that is shared or easily communicated to the other partner. Therefore, in a permanent marriage we would expect both a similar number and a similar quality of M responses. Conversely, persons with very dissimilar M responses are expected to be incompatible in the sense that they are not able to communicate with and understand each other on a deep level, their levels of psychic complexity, both emotional and intellectual, being too disparate.

The level of complexity can be defined as the degree of psychic awareness, subtlety of thought, introspectiveness and creative fantasy which a person develops. In other words, it is the degree of individuation and ego integration achieved by a person, the essence of his uniqueness as an individual. We assume that the quantity of M in a Rorschach record reflects the degree of complexity of any person's self-projection into the world, while the quality of the M reflects the manner in which this complexity is shown or used.

An individual of average intelligence produces 2 M. Individuals of superior intelligence, with intelligence quotients of at least 110, produce about 5 M on the average. There is, then, some positive correlation between the number of M and intellectual level, but it is not high.

Another factor that influences the M at least as much as intelligence is interest in interhuman relationships. Thus, for example, physicists, paleontologists and others concerned mainly with the study of inanimate matter have, on the average, significantly fewer M than do artists, clinical psychologists, psychiatrists, psychiatric social workers and the like who are primarily occupied with interhuman relations.

On the whole, our patient groups were of superior intelligence and their M numbered more than 2 on the average. The essential point, however, is that marital mates with a difference of 3 or more in their sums of M would tend to manifest such disparate levels of psychological complexity that permanent and real understanding would be impossible between them. Also, their mental relations would most likely be superficial and imperfect.

In order to verify this conclusion, we compared the number of M in two groups of patients. One group consisted of couples who had been divorced, and the other of pairs who continued to live together despite serious dissatisfaction with the marriage on the part of both husband and wife. All of these patients had sought marital advice primarily, needing help in comprehending the causes of their marital difficulties. Their Rorschach records were obtained in connection with the marital consultation.* A comparison of the records of happily married couples, matched in various important traits, with the findings in the other two groups, would have been ideal. However, such records could not be obtained.

The main objective difference between our two groups was divorce. In both groups there was open conflict, consideration of separation or divorce, little compatibility and marked unhappiness. Yet the couples in one group continued living together as man and wife despite the tensions and misunderstandings, while those in the other had been divorced.

Because of the difference in number, the group of divorced couples could not be matched perfectly with the group of unhappily married couples who were continuing their marriages. There were twenty-two divorced couples and thirty-three unhappily married pairs. This numerical discrepancy caused us to create a third group by matching at random men and women in the unhappily married group of couples who maintained their marriages. Our assumption was that randomly matched couples (randomly matched Rorschach records of a man and woman)

---

* The authors wish to thank Mrs. Sophie Gottlieb for the use of many of these Rorschach records.

would not show a fundamental compatibility. We took the husband of one marriage and paired him with the wife of another marriage, matching each woman as to educational level and age with the real wife for whom she substituted.

Of the total of fifty-five couples studied, the duration of marriage at the time of the Rorschach examination ranged from two to forty years. Nineteen couples had been married more than sixteen years, seven couples less than six years, and one-third had been married from six to ten years. The number of children varied from none to three, with 42 per cent of the couples having two children and 18 per cent none.

The husbands' chronological ages were divided as follows: three were over fifty years old (two in the divorced and one in the unhappily married group); seventeen were between forty-one and fifty; one (whose divorced wife was of the same age) was below twenty-five. The largest number—34 or 62 per cent of the whole group—were between twenty-five and forty years of age.

The educational level of the husband decisively determined the socio-economic level. Three per cent of the husbands had only a public school education, 37 per cent graduated from high school, 24 per cent had a college education and 36 per cent had professional graduate university training. Most of the subjects came from an urban environment (New York City), were of the same religious faith (Jewish), and were eager to be successful in life both financially and socially.

Obviously, in isolating M as a factor to study we were dealing with only one of the components which enter into the complete interpretation of the Rorschach record. It cannot be denied that each marriage is unique and that what is crucial to the happiness of one may be of no consequence in another. Therefore we could not hope to find compatibility in terms of the M in 100 per cent of the married couples who continued living together despite discord. However, it was to be expected that this one level of deep relating should be significantly higher in the maintained marriages than in those which had been terminated.

TABLE I—*Difference in absolute numbers between husband's and wife's Rorschach human movement responses in three subject groups: unhappily married couples, randomly matched couples, and divorced couples.*

SUBJECT GROUPS	ABSOLUTE DIFFERENCE IN NUMBERS OF M								AVERAGE GROUP DIFFERENCE
	0	1	2	3	4	5	6	OVER 6	
A. 33 married couples	4	10	6	3	4	2	3	1	2.58
B. 33 randomly paired couples	1	2	5	6	7	4	5	3	4.21
C. 22 divorced couples	1	2	4	3	4	3	2	3	4.09
All 88 couples	6	14	15	12	15	9	10	7	3.57

Table I shows the distribution of the difference in the number of M between husband and wife. It should be borne in mind that Group B consists of members of Group A, but that the couples in Group B were randomly paired to test the assumption that similarity in the number of M is associated with mating in marriage and that consequently theoretically mated couples would not exhibit the same degree of similarity in the number of their M as do actually married couples. Group C constitutes the second control group. Being divorced, these twenty-two couples were expected to manifest a lower degree of similarity in their numbers of M than in the still married couples.

So far as our patient groups are concerned, it does not matter at what exact point the difference in the number of M is considered large and significant. Whether we set the critical point at 2 or 4, the sum of married couples with low differences in M and divorced couples with large differences in M is about the same. In view of the imperfect reliability in scoring the M, it seemed advisable to consider the difference in the absolute numbers of the husband's and wife's M as large and significant when it exceeds 3, rather than 2 points.

The statistical chi-square test shows the association between differences in the husband's and wife's numbers of the Rorschach human movement responses and duration of marriage to be significant with a $P < .01$. This measure implies that such an association could have occurred by mere chance less than once in one hundred investigations and that, therefore, this association between permanence of marriage and differences in M below or about the critical point (difference of no more than 3 points versus difference greater than 3 points) would be confirmed in future investigations conducted in essentially the same manner as this one. However, it does not imply that the degree of association, though significant, will be found to be exactly the same in re-investigations.

When Group A (married couples) was compared with Group B (randomly paired couples), it was found that in twenty-three married couples the difference in the numbers of M was small and in nineteen randomly paired couples this difference was large. These two subgroups constituted 66.7 per cent of the total of sixty-six couples in Groups A and B.

When Group A was compared with Group C (divorced couples), the twenty-three married couples with small differences in the numbers of M plus the twelve divorced couples with large differences in the numbers of M constituted 63.6 per cent of the total of fifty-five couples in Groups A and C.

A positive association between small differences in M and continuance of marriage on the one hand, and between large differences in M and divorce on the other, was found in 64 to 67 per cent of the cases studied. This association can also be expressed in other statistical terms: the percentage of couples with small differences in the husband's and wife's numbers of M (not more than 3 points) was 69.7 in the married though unhappy Group A, 42.4 in the randomly paired couples of Group B, and 45.5 per cent in the divorced Group C. The difference in the percentage of Group A and the percentages of Groups B and C is significant statistically.

For better results, it may become necessary to relate the difference in the numbers of M to the absolute quantities of M. The same numerical difference in the M may indicate a greater difference in the pscyhological complexity of the marital partners when their M are few than when their M are numerous; a difference between o and 4 is probably more significant than one between 10 and 14. A larger number of cases and longer and more detailed observations and examinations than we were able to make are necessary to refine the relation between the size of the difference and the degree of difference in psychological complexity between husband and wife.

Although the findings for the M are significant at a high level of statistically determined confidence, the wide overlap between the unhappily married couples on the one hand, and the randomly paired and divorced couples on the other, indicates that the psychological factors implied by differences in the number of M cannot be the sole or sufficient cause for terminating a marriage. Other prominent causal factors must be assumed. However, it is certain that the degree of similarity in psychological complexity, as revealed by the number of M, is one of significant and important factors determining the duration of marriage.

Up to this point, only the quantity of the M has been discussed. Other important Rorschach components are the quality of the M and the degrees of similarity among M of different qualities. If the quantity of the M shows how much psychological complexity there is, the quality of the M reflects the manner in which the complexity is expressed, the kind of role in life a person has adopted for reasons usually unknown to him and certainly beyond his conscious and easy control. Frequently the individual is not even aware of the role he is playing when dealing with others in personally vital matters.

Varying greatly, the roles can be assertive, aggressive, cooperative, compliant, passive, ambivalent, doubting, etc. Some can be played by only one marital partner if the marriage is to be harmonious, while some can be shared by both partners without endangering the marriage. Our study showed that in the great majority of both the married and the divorced couples the M of both husband and wife were of similar quality. However, the married couples' M, though similar, tended to be less similar than the M of the divorced couples.

There was more complementation in the records of the married couples, an understandable factor. If both partners were aggressive to the same extent and tended to express it in the same manner and under the same circumstances, there would be too many quarrels and basic disagreements for the marriage to work. It is easier to get along if the desire to win out in competition or in neurotic defensive aggression is weak in both partners, or at least in one of them. This aspect of the study requires a much more subtle analysis and a larger number of cases to confirm the findings.

Chromatic color responses are Rorschach's second crucial factor in revealing the individual's way of life. The color responses indicate the intensity and quality of the desire to associate with or dissociate from others, with the intent of a voluntary or forcible continuance or discontinuance of exchange of pleasures or pains with them.[1] These needs and the CR which represent them in the Rorschach method are more variable than the M and the traits indicated by the M.[2] Lack of color responses indicates lack of need for direct and close emotional experiences with others.

When we compared our divorced couples with unhappily married pairs who nevertheless were determined to keep on living together, we were unable to discover any statistically significant difference in terms of the ratio of the sum of M to the sum of CR. As to the sum of weighted color responses, there was practically no similarity between husband and wife in any of

the three subject groups. In fact, no other Rorschach component manifested such a conspicuous lack of similarity between
husband and wife.

In regard to the quality of color responses, many husbands
produced superficial and shallow color responses (nature or un-
derseas scenes, colored objects) while their wives (especially in
the divorced group) produced earthy or intense color responses
(leaping flames or fire). This difference may explain in part the
neurotic attraction which made the marriage possible, but failed
to maintain it.

We can speculate that perhaps the shallow affect of the husband would thrive on the hectic intensity of his wife, borrow-
ing a little life, so to speak. Conversely, the emotional fire of the
wife might be toned down by the lack of emotional intensity in
the husband, affording her some external control and protection against her own impulsiveness and intensity. At the same
time, however, there was constant frustration and accumulation
of explosive and aggressive energy, increasing the incompatibility between the marital partners. Neither partner obtained an
adequate emotional response from the other.

Color responses, such as "orchid" or "cross-section of a
flower," projected into plate IX were particularly frequent in
the records of the divorced husbands. The divorced men produced a larger percentage of "delicate" visual images than the
non-divorced ones.

In line with this problem is the consideration of the incidence of M-shock in the records of the divorced women. An
M-shock[1] is inferred from the inability to give a human movement response to plate III, which elicits M more easily than any
other of the Rorschach blots, particularly if M was elicited by
other blots.

Five of the divorced women failed to produce the popularly
seen M in the dark areas of plate III, although they had M elsewhere in the record. Five other divorced women projected non-
human figures into the dark areas of plate III; they endowed

them with human characteristics and engaged them in movement or made them assume a posture, thus creating a dehumanized M.

M-shock can be inferred even if there is no M in the entire record, provided the subject displays signs of hesitation, delay and obvious increase in tension when shown plate III. Nearly half of the divorced women—ten out of twenty-two—had an M-shock, a shock reaction which is rare and which was much less pronounced in the records of those who maintained their marriage despite the unhappiness and conflicts. The M-shock indicates neurotic ambivalence concerning the life role that the individual is unconsciously prompted to play.

Tentatively, then, we can infer that the divorced women tend to display, more frequently than do married women, a deep problem in relating to others, probably to men in particular. This may be associated with insufficient identification with the female sex role, or with an inability to assume an adult female social role that involves a responsible relationship.

In conclusion, it remains to be seen whether the findings of this investigation apply to other subject groups or are specific to the big city (or metropolitan) group, composed of rather energetic people eager to succeed in a highly competitive society. These factors cannot be disregarded in any evaluation of the validity and applicability of our conclusions.

This investigation has used the Rorschach technique as a tool of analysis, concentrating on one aspect of the method, the human movement response. We realize that the variables determining the duration of marriage are many and manifold. It is hoped that our findings, which suggest new areas of intensive analysis in marriage research, will be confirmed by investigations of other population samples.

References

1. Piotrowski, Z. A., *A Rorschach Compendium*. Revised and enlarged. Utica, New York, State Hospitals Press, 1950.

2. Piotrowski, Z. A. and Schreiber, M., "Rorschach Perceptanalytic Measurement of Personality Changes During and After Intensive Psychoanalytically Oriented Psychotherapy." *Specialized Techniques in Psychotherapy*. New York, Basic Books, Inc., 1952.

3. Rorschach, H., *Psychodiagnostics: A Diagnostic Test Based on Perception*. Berne, Switzerland, H. Huber, 1921.

Changes in Family Equilibrium Through Psychoanalytic Treatment

VICTOR H. ROSEN, M.D.*

IN MANY RESPECTS the family relationship resembles what physicists call a "closed energy system." By law and tradition it is one of the most protected relationships existing in civilized society. Indeed, the family is a law unto itself, a microcosmic state within the state. It can legislate its own rules and codes which, subject only to certain broad limitations, are immune from outside authority. Complex interpersonal adaptations and accommodations become implicit in the individual family structure during its growth and evolution.

The most striking single factor in the analysis of a married

* Assistant Medical Director, Treatment Center, New York Psychoanalytic Institute; Assistant Attending Psychiatrist, Mount Sinai Hospital, New York City.

patient, or of a child who is part of a functioning family unit, is the introduction of a new person—the analyst—into the previously closed intimate circle. Unlike the marriage counselor, lawyer, family physician or clergyman, this new partner is not only admitted into the private life of the family, but also into that part of their relationship that they have managed to keep hidden from each other and even from themselves. The prognosis of the family relationship is often bound up not only with the effect of the analyst on the psychic economy of the patient and the alterations in the family equilibrium so induced, but also with the unconscious meaning of the analyst and the treatment for the untreated members of the family group.

Transference problems of the untreated marital partner are a tenuous and little considered aspect of psychotherapeutic relationships because of the difficulty in obtaining accurate information. Unless other members of the family are also in treatment, concurrently or otherwise, or are seen in consultation, most of the data must be gathered indirectly from the patient, and is therefore subject to considerable distortion. Frequently, however, there are direct or indirect communications from other members of the family group which give important glimpses into their feelings about the patient's therapist and treatment.

Most transference reactions are wholly or partly covered by a façade of reality. Despite its use as a façade, however, this reality must be given the recognition that is its due if underlying irrational elements are to be uncovered. This is as true in understanding the reactions of the untreated members of the family as it is for the patient.

Inevitably, the untreated members of a closely knit family group have a large and important stake in the treatment of the patient. Upon the success or the failure of the therapy may hinge such critical issues as divorce, remarriage, child bearing and the economic security of the family. The very seriousness of these considerations, even in the most stable individual, is bound to reawaken certain unconscious patterns of dealing with anxiety. These will stem largely from infantile sources and may be directed toward the analyst into whose hands so much has been entrusted.

Here it would be well to indicate briefly the source of such infantile patterns. Although the anxiety of the relatives is mobilized by the realistic considerations of the treatment, the unconscious component of the reactions, which can be described as "transference phenomena," arise by and large from the existence of the new "triangular" or "polygonal" situation itself. In general, this component is derived from two familiar triangles of childhood experience—the oedipal relationship, and the sibling rivalry struggle for one or the other parent.

The deep psychological need to reconstruct the childhood constellation, even in certain aspects of normal psychology, is discussed by Helene Deutsch in describing feminine eroticism and motherhood: "It is erroneous to say that the little girl gives up her first mother relation in favor of the father. She only gradually draws him into the alliance, develops from the mother-child exclusiveness toward the triangular parent-child relation and continues the latter, just as she does the former, although in a weaker and less elemental form all of her life. . . . While waiting for motherhood, even before its beginning, woman psychologically prepares for the triangle. Sometimes this is expressed directly and consciously in the wish 'I want to have a child by him' . . . at other times the wish may be 'I want to have a child' and then the man is partly moved into the background. . . . The normal woman always more or less includes the man, and this not only in the deep physical sense, because the formation of the triangle is a deep need for her. This need often asserts itself under the most unexpected conditions, and the failure to satisfy it can considerably disturb her relationship to her child." [1]

Even before the treatment of the patient has actually begun, certain general signs of this reaction are often discernible.

While there are many variations, the following are some of the characteristic attitudes:

1. The analyst is viewed as a rival for the patient's affection or loyalty. This may lead to overt or covert attempts to delay or disrupt the treatment.

2. The analyst is felt to be an ally. Attempts are made to communicate grievances and advice to him, to seek his aid on redresses for wrongs suffered at the hands of the patient, to advise him on technical procedures on the basis of long experience with the patient's idiosyncrasies, or to ask advice on how to handle the patient so as to be of service to the analyst, etc.

3. The analyst is seen as an impotent figure. The patient's lack of progress is emphasized, and unwillingness to delay irreversible decisions or to harbor any optimistic expectations for the treatment is a likely concomitant.

4. The analyst is viewed as omnipotent, as one who has limitless capacities and knowledge. The family relationship may settle down into a caricature of security and contentment with many immediate issues left unfaced or not acted upon.

5. The therapist is regarded as a punitive and depriving figure against whom the family seeks to form an alliance with the patient, with the ultimate defeat of the analyst as the unconscious aim.

6. The analyst is idealized and the need for his admiration is emphasized by a bitter internecine struggle between group members to assure themselves of his favorable judgment.

Generally, there are three sources of data in regard to this problem. First, there is the insight gained from the revelations of the patient himself in the course of treatment. Secondly, there is the understanding gained through subsequent treatment of other members of the family either by the same or a different therapist. Thirdly, consultation with other members of the family by the patient's analyst or another therapist is often helpful in yielding information.

The following clinical example is a case in point: Mrs. Y., a thirty-three year-old woman, was seen in consultation for several interviews. Her husband was currently in analysis. A successful professional man, he was apparently somewhat inadequate as a husband and father. Referring to some psychotherapy she had had four years previously, Mrs. Y. stated that since then she had been able to resolve a "career versus motherhood conflict" and had given birth to two more children. The couple now had five children. Mrs. Y. had been fairly content until her husband began his treatment—a step she had constantly urged upon him. As her husband's treatment progressed, she started developing "unreasonable fears" that her husband would desert her. She interpreted each family squabble as a threat of divorce. When her husband arrived home late she berated him, though she knew he had just come from his analytic session.

An intelligent and sensitive woman, Mrs. Y. revealed a considerable capacity for condensed insight into her less apparent feelings and motivations. It became clear that she carried within her a fantasy of her husband's analyst as a diabolical manipulator and an omnipotent being. She felt he was interested only in his patient, and that through his analysis he was going to reveal her inadequacies to her husband. Her husband's cure would mean his becoming aware of those defects in her of which he had previously been blissfully ignorant. This could only terminate in a divorce that would leave her alone in her middle-age with five children to care for and no chance to remarry.

It was suggested to her that she somehow conceived of herself as a helpless bystander whose future was being plotted by another person and that her own consent or refusal to engage in divorce proceedings did not seem to enter the picture. She then gave an account of a painful incident in her childhood. When she was four years old, her mother had become ill with what proved to be a chronic disease requiring hospitalization. There were several older siblings, all of whom were already going to school. The patient was dimly aware at the time that some sort of plans were afoot to send her to live with her grandparents in another city.

This separation from the family lasted several months. She believed it was her father who took her to the grandparents' home. As an adult, the patient has reproached her parents for this separation. Although she realized that the next oldest sibling, a brother three years her senior, was kept at home because of his schooling, she still felt the brother should have been sent along for companionship so that the temporary absence from the family would have been less painful. Until the present discussion, she was not aware that it was really her father whom she blamed most severely for what had seemed at the time to be an act of desertion.

It is most unusual to be able to reconstruct the significant childhood episodes which are prototypic of the transference on such brief contact with a patient. In this case previous psychotherapy had apparently prepared the ground, making it possible to have more direct access to the pertinent data. It is of interest that this particular childhood event with its many meanings had been discussed in the patient's previous treatment but in a totally different light.

In the current situation, her husband's high regard for his analyst, his unwillingness to discuss his analysis with her, and her feeling of being excluded from any participation in its outcome, were all stimuli to her feelings. The important factor to note here is that the threat of desertion was derived not from current circumstances but from the past event in this patient's life.

Another case illustrates a similar problem in somewhat different form. Miss Q., an adolescent girl of nineteen, had been under analysis for about eight months when it suddenly reached an impasse. Her father, who had been paying for the treatment, now demanded that a time limit be set for its termi-

nation. In a three-way interview, necessitated by his insistence
and the patient's fear that he would interfere, several matters
became clear. The patient's homosexuality, which had never
been made known explicitly to the father, was apparently
sensed by him and felt as a reproach for his own limitations as a
parent. Just about this time, the patient had gone out on her
first date with a young man, an event that seemed to awaken
something in the father and motivated his interference.

Ostensibly, he wanted his daughter to be able to fall in love
and marry. During the interview, however, it was obvious that
he felt hostile and highly competitive toward the analyst who
was treating his daughter. Speaking of himself as a "self-made
man," he repeated over and over that none of his daughter's
problems, whatever they might be, stemmed from him. Had she
been able to confide in him, as a good daughter should, he
would have been perfectly capable of giving her the necessary
guidance to solve her problems, and there would have been no
need for psychoanalysis.

His attitude of competitive rivalry in parenthood, anger at
the analyst, and resentment at having his role apparently
usurped, continued throughout the treatment. During the interview, and from what was learned later in the patent's analysis, it seemed likely that the father was re-enacting his own early
oedipal conflict and that, despite a certain sophistication, he
could not tolerate the feeling that another man was ostensibly
outdoing him in his parental role.

During psychoanalytic treatment the family equilibrium
can be tilted in either favorable or unfavorable directions. These
are too numerous and too uniquely individual for more than a
cursory discussion. Changes can occur in this equilibrium not
only during the course of the analysis but also after it is terminated. The fact that the family group tends to be the most accessible stage for the re-enactment of transference phenomena
makes it inevitable that more or less observable changes in the

balance of forces will become apparent during the progress of most analyses. Transference phenomena are seen in their most direct forms when growing children repeat earlier stages of their development in relation to their parents. This is more subtle but no less discernible in parents who react to each other or to their children in accordance with more primitive biographic stages of their own development.

The most dramatic changes in family equilibrium are observed in cases where "acting out" is a significant defense of the patient. To many laymen, "acting out" within the family circle is still regarded as the effect of the psychoanalysis itself. Many combinations and permutations of this relationship are possible. An analysis may activate the process of "acting out" in one or more of three directions: toward the analyst in the analytic situation itself; as an intensification or abrupt change of a certain relationship within the family; or it may discharge itself in certain kinds of behavior and activity directed toward the outside world.

To the extent that the patient's conflicts and childhood regressions are focused on the analyst, the family group or the outside world are given some respite from tension and the resultant equilibrium, though temporary, is a favorable one. An apparent improvement and calm may settle over the household. But to the degree that these energies are concentrated upon the members of the family group, the reverse holds true. Sometimes, when the "acting out" is directed toward people outside of the family group, certain other problems may be introduced which likewise shift the inner group economy.

As other defenses in the patient's psychic economy become strengthened or modified, similar but more subtle rearrangements may often become manifest. For example, a young married man in analysis, who was beginning to understand the role played by his passive feminine drives in the stormy relationship he had with his wife, reported that for some reason his wife and their seven-year-old daughter had suddenly become good friends.

Previously the mother-daughter relationship had been described as "a constant bickering back and forth, just short of hair-pulling."

What the patient had not observed at this time was that his own attitude toward his wife had changed considerably. Now, instead of taking every opportunity to belittle her and prove he could be a better "mother" than she, he found himself supporting his wife in the justifiable disciplinary measures she imposed on the child. The new respect acquired by the wife in his own eyes increased her standing with the daughter as well.

Readaptive stresses in a family group are particularly apparent when some radical alteration in character structure has occurred in a patient who has undergone a successful analysis. These may become particularly acute in the period following the termination of the treatment. In regard to marital partners, especially, the unconscious factors in the choice of the partner may have had much to do with his or her psychopathology. It is not infrequently found, for example, that women suffering from various forms of frigidity choose male partners suffering from potency problems.

In such a marriage, the restoration of potency to the husband by a successful analysis will almost inevitably produce a readaptational pressure on the wife which increases with time. This new equilibrium may take place in a variety of ways. The frigid wife may spontaneously overcome the sexual symptom, if it is not too deeply rooted, in response to the change in the patient. Or the situation may become intolerable to one or both partners and lead to divorce. Or the frigid partner may for the first time become aware of her symptom and seek treatment for it. The same shifting accommodations can be observed in more subtle forms where the alterations in the patient at the termination of an analysis are less obvious in kind and intensity.

Writing on the concurrent treatment of married couples elsewhere in this volume, Mittelmann indicates that there may be favorable and unfavorable readaptations so far as the marital relationship is concerned.* He stresses the transference pitfalls in this procedure, but also cites what he feels to be its advantages over separate analyses.

Out of twelve such pairs reported by Mittelmann in a previous paper,[2] the husband and wife of four couples had daily analytic interviews over prolonged periods of time. Two of these couples had terminated their analyses, with divorce resulting in one case, and an apparently improved relationship in the other. In eight instances one of the mates came for analysis, while the other received a minimum of two but not more than twenty interviews at intervals of one week to several months. In six of these couples, there was sufficient change in the mates' reactions to make the relationship satisfactory for the patient. Three of this group had terminated their analyses successfully at the time the report was published. A similar study has also been reported by Oberndorf.[3]

In many so-called "sado-masochistic" marital relationships the process of attack and counterattack becomes so intense that it constantly threatens the partnership. These are among the most frequent marital patterns in which both partners may seek analysis. The very fact that the individuals concerned feel that each may be contributing to the hostilities is in itself a big step toward constructive altering of the explosive situation.

In dealing with such problems the analyst must be on the alert for unexpected and devious use of marital flare-ups for purposes of "resistance." Such patients often attempt to turn the analysis into a court of domestic relations. Righteous indignation at the wrongs suffered at the hands of the partner, like radio "jamming" of propaganda broadcasts, has the effect of interfering with the communications of the analyst. It is during a lull in the battle, sometimes produced by a significant step in the analysis of one of the partners, that important insight can become available to the other.

---

* See Chapter VI.

An example is the case of a capable young man with an unconscious tendency to interfere with his own success, anxieties about his work, and a growing and mutual exasperation with his equally insecure wife. For several years the marriage had been a constant battleground. Each of the partners felt misunderstood and cruelly treated by the other. After the husband entered analysis, it soon became apparent that the wife's difficulties were of sufficient intensity to warrant treatment for her as well. Clearly, both individuals had problems in regard to self-esteem, and the slightest hint of criticism on the part of one brought immediate reprisals from the other, with a pyramiding effect.

The dynamics of the husband's problem lay hidden in his ambivalent relationship with an ambitious and successful father. In his work inhibition the young man succeeded in reenacting and disguising this problem. His aggressive and competitive attitudes toward other males, and his fears of retribution from them should he be successful, were largely unknown to him. Because of this anxiety, he found no self-assurance in his manifest capabilities. While he considered his wife's sarcastic comments as the instrument producing his wounds, they were actually merely the salt which was being rubbed into them.

The patient was only aware of regard and respect toward the analyst. All efforts to uncover evidence of rivalry and fear in this relationship were successfully eluded by well-timed brawls between the marital partners. On one occasion, following an attempt to delineate this aspect of the transference situation, the patient returned home to his wife and tried to start a quarrel in the usual way. However, for once his wife refused to enter into battle and instead became playfully affectionate. That night the husband had a nightmare which clearly portrayed the analyst (father) as the major antagonist.

Subsequent analysis of the patient's conflicts over competition succeeded in increasing his work capacities and self-esteem, making it possible for him to be less vulnerable to the usual marital tensions. This highly condensed version of the interacting forces and intricacies of this case may suffice to illustrate the effect of treatment in altering marital equilibrium, an experience that is not unusual in analytic practice.

In a procedure as highly personal and individual as psychoanalysis, any statistics on the results of one aspect of the patient's treatment, namely the outcome of his marriage, would have little meaning in themselves. Not every divorce means failure of the treatment in a particular case. Nor can the tightening of a marital bond in itself be taken as an indication of therapeutic success. In addition, there are the difficulties in quantifying and characterizing individual treatment procedures.

For these reasons, the effects of psychoanalytic treatment on the family group cannot be measured with scientific accuracy, but must remain matters of expert conjecture and intuition provided by experience. There is no doubt, however, in the opinion of many patients as well as analysts, that the relationship with the marital partner and the children takes on a new and better meaning after unconscious conflicts in one or more members of the family have been resolved.

The variety of relationships in a family group where one or more members is under psychoanalytic treatment might be roughly schematized in the following way: The lines drawn between analyst and patient, and patient and various members of the family group could be clearly defined and heavily drawn. Those connecting family members and analyst would be drawn as faintly dotted lines. As treatment progresses, the dotted lines consolidate by virtue of the transference forces operative within the analysis. The forces of interaction between the family members are thereby modified.

Such a diagram facilitates the examination of several aspects of treatment such as external interferences with the course of the treatment, the character of the relationship between the patient and others in the family group, and some of the pertinent emotional problems of the untreated members of the family.

Detailed study of the subtleties of this problem will shed light on such questions as these: What is the effect of the analyst's interviews with untreated family members upon the patient in treatment? What is the best timing for suggestions concerning consultation with family members? What countertransference problems appear as a result of these contacts? It is hoped that a study of these problems will add to the theoretical and technical range of psychoanalytic knowledge.

References

1. Deutsch, Helene, *The Psychology of Women*, Vol. I. New York, Grune and Stratton, 1944.
2. Mittelmann, Bela, "The Concurrent Analysis of Married Couples." *The Psychoanalytic Quarterly*, XVII, 182, 1948.
3. Oberndorf, C. P., "Psychoanalysis of Married Couples." *Psychoanalytic Review*, XXV, 453, 1938.

General References

Fluegel, J. C., *The Psycho-Analytic Study of the Family*. London, Hogarth Press and The Institute of Psycho-Analysis, 1929.
Kubie, L. S., *Practical and Theoretical Aspects of Psychoanalysis*. New York, International Universities Press, 1950.

The Problem of Family Diagnosis

MARCEL HEIMAN, M.D.*

ANY APPRAISAL of the family scene of today, without adequate knowledge of its gradual development over many thousands of years, may be compared to an attempt to arrive at a critical opinion of a painting after viewing it for a fleeting instant through a tachistoscope. One can hardly see the painting, let alone grasp its meaning.

From the earliest days of man's history, marriage, in addition to being the beginning of the family, has also had its biological and psychological roots in the family. Obviously, then, a great deal may be learned regarding the harmony and disharmony of present day marital partners by studying the infancy and childhood of each within the context of their individual family settings.

Historically, psychoanalysis has confined itself to the individual and his intrapsychic processes and conflicts. However,

---

* Assistant Attending Psychiatrist, Mount Sinai Hospital, New York City.

Freud's discovery of the pervasive influence of the Oedipus complex on human life, together with his "Contributions to the Psychology of Love," laid the groundwork for a psychoanalytic study of marriage.[5]

Fluegel, in his classic volume, *The Psychoanalytic Study of the Family*, amplified our knowledge of the dynamics of family life. "It would seem probable indeed," he stated, "that a thorough understanding of the problem of love, sex and marriage cannot be attained without a preliminary knowledge of the nature of the psychic bonds that unite parent and child—a knowledge that psychology is only now beginning to afford."[4]

These observations, made more than thirty years ago, may now be brought up to date with the recent contributions on ego psychology, the pre-oedipal period of development and the psychodynamics of interaction. In the light of psychoanalytic understanding of parent-child relationships and early childhood experiences, longitudinal studies of individuals from childhood through adulthood into marriage furnish the necessary insight into the dynamic continuum that stretches from birth and early infancy to later life, connecting past and future generations.

Family tensions in modern life imperil the very foundation of our society. So far, we have been able to withstand the attempted perversion of the family by the Nazis and the efforts at weakening family ties by Communists. But in addition to these destructive forces there are intrinsic ailments in the family that sap its vital strength as a social unit, as well as the strength of the marriage on which it is based. We have not been sufficiently alerted to the fact that an unhappy marriage is an illness for which we should seek medical aid. To quote Ernest Jones, "An unhappy marriage (and somatic illness) holds a neurosis at bay."[6] In other words, an unhappy marriage may even satisfy the unconscious sense of guilt, and improve a neurosis.

Psychoanalytic investigation of marital partners is beginning to help us to understand the dynamics of this basic social institution. If we assume that in marriage we shall find a repetition of important experiences of the past, especially the two most significant ones—the mother-child relationship and the oedipal relationship—then in order to understand all the conscious and unconscious motivations and satisfactions of a marriage, we must trace the lives of the marital partners from birth into marriage.

The mother-child relationship, the oedipal complex and marriage, in their ever-repeating cycle and their interlocking, represent nodal points or stations at which civilization manifests itself, among other things, as a neurotc expression. It is as part of this cycle that yesterday's neurosis becomes one of the accepted life experiences of tomorrow.

The choice of a marital partner in founding a family is but one point in the continuous line that starts with birth and ends with death. Sometimes two young people meet for the first time, and the moment they lay eyes on each other fall in love. This remarkable occurrence is known as "love at first sight." The dazed pair ascribe it to the fact that "fate brought us together." Or they say, "it was written in the stars." Such phrases imply that a superhuman power was responsible for the phenomenon. It is easy to see that fate in these cases is the projection of an inner force of which its carrier is quite unaware. Just as no physical body could refuse to obey the law of gravitation, so no human body can desist from acting according to the forces within his own unconscious which propel him along a certain course.

An understanding of marriage and the choice of marital partner must be preceded by proper observation of the courtship and a prior understanding of teen-age and adolescent development and, most important, of early childhood and infancy. Concurrently, during all these phases, the development of the individual has to be observed within the framework of his relationship with important members of his family, especially his parents and their substitutes.

Freud once referred to the mother-child relationship as the "smallest group." With the mother-child relationship being fundamental for every individual, marriage becomes the universal repetition-compulsion of mankind. Therefore, as has been noted, it follows that much of what we have learned and will learn about the mother-child relationship will find application in our examination of another "small group"—the marital relationship.

In the field of emotional interaction certain aspects of the mother-child relationship are paralleled in the marital situation. Dorothy Burlingham speaks of "direct communication" between the unconscious of the mother and that of the child.[2] She refers, for example, to a sado-masochistic relationship between mother and child, with each provoking the other—a *"folie à deux."* I have made similar observations in married couples, whose reactions were exposed in the course of analysis, as mutual unconscious provocations.

The trained observer, not to mention the layman, finds it difficult to understand the unconscious communication and interaction between marital partners. One reason is that some of this communication is non-verbal, or, as in the case of the child, pre-verbal. There is a tendency, completely unjustified, to relegate this form of communication to the realm of the mystical or telepathic. Acutally, there is an affective communication between marital partners as there is between mother and very young child, and man and animal.

Consider the possibilities if A's unconscious wish is communicated to B, without either one being aware of it; is then acted upon by B, and finally consciously reacted to by A, who started the whole thing in the first place. This is marital interaction in action.

For example, a woman patient had strong scoptophilic tendencies. Since these would have been most disturbing had they emerged on the surface, she had to repress them. She could maintain this state of affairs only if her husband peered into neighboring windows through his binoculars for hours on end. Gradually it became apparent that she was unconsciously prompting her husband to indulge in this activity, while consciously criticizing him for it.

This situation is reminiscent of an incident told me by a married couple. Having purchased two electric blankets for their twin beds, the pair with the help of their two children set about hooking up the various electrical wires so that the blankets would be ready for use that night. The control gadgets were placed on the night tables alongside each bed. When the couple retired, the husband was unable to fall asleep. He found the blanket much too warm and turned the heat down on his regulator. Meanwhile, his wife felt rather chilly and turned the heat up on her regulator. But the husband still felt too warm and turned the control down a bit more. At the same time his wife, feeling colder than ever, turned on still more heat.

This went on until he could not stand the heat another minute, while she was freezing to death. It was at this point that they discovered what was wrong. It seems the wires to their blankets had somehow gotten crossed. Similarly, all manner of wires get crossed in the constant and subtle emotional interaction between man and wife.

Because a marriage is based on a number of conscious and unconscious motivations and expectations, the choice of partner is quite important, as is the manner in which that choice is made. Attention has recently been focused on this aspect by Henry V. Dicks and Nathan W. Ackerman. Dick stresses the "importance of role expectation." [3] Ackerman points out that each of the marital partners, seeking in the other the love and protection of a parental figure, pushes the relationship toward the needed child-parent form. He calls this attitude an attempt to "parentify" each other.[1] While this "role expectation" is frequently encountered, it should not be oversimplified. By the very nature of our life experiences, we expect the marital partner to play not one role but a variety of them. Some of these roles will be of greater importance than others, some may be directed toward the same goal, others may have opposing aims.

The psychoanalytic study of such role expectancies enables us to understand their instinctual source, as well as the specific defense employed to keep such instinctual drives in check. Marriage is the stage on which the individual plays multiple roles and expects his mate to play multiple roles too. A woman, for example, chooses her husband because she sees him as a carbon copy of her father whom she disliked. This permits her to reject her husband, a role that becomes permanent on her part. Her husband, in response to her rejection of him, frustrates her, thereby fulfilling the expected role of rejecting her as her father once did.

What is more, their relationship permits the wife to find expression for her ancient masochistic needs in relation to her mother. Further, the wife's ambivalence toward her mother figure is encouraged by the fact that her husband had two mothers: a real mother whom he loved and who died, and a stepmother who rejected him and whom he hated.

Furthermore, in the unconscious the marital partner may assume other roles, for example the super-ego, also of one's ego, one's id and one's various defenses.

Obviously, for our purposes it is not enough to concentrate on the unstable marriage and on those forces that bring about its breakdown. Equal attention must be directed to the stable marriage and those forces that maintain it, particularly because such stable marriages seem to contain compensatory mechanisms for neurotic or psychotic individuals. One marriage may be maintained successfully *because* of a symbiotic relationship between husband and wife. Another union may be broken in short order because the symbiotic need of one partner cannot be met by the other. An understanding of the family and the possibilities for helping them would require not only a psychiatric diagnosis of each member but a psychodynamic evaluation based on the interplay of unconscious attitudes and emotions.

The Smith family is a vivid illustration of neurotic interaction, with its inherent frustrations and hidden gratifications.

The Smiths, both in their thirties and college graduates, came to an agency for help with a rather serious marital problem. Mrs. Smith was apparently incapable of doing any housework. The situation had become so critical that soiled dishes and clothes had piled up for weeks. She neglected the children's cleanliness and appearance as well as her own.

Mr. Smith had tried for a long time to cope with this state of affairs by directing all of her activities during the day. He would make detailed lists of what she was to do, where she was to go, and the like. He even went so far as to tell her what clothes to wear each day, what kind of dresses to buy, how to have her hair done at the beauty parlor, and to give her a host of other instructions that were meant to be helpful.

The couple have been married for ten years and have two children, both of whom have developed neurotic traits. The boy of nine is self-centered, whines and is frightened of being alone. The seven-year-old is a bed-wetter, sucks her thumb and is given to temper tantrums.

Because Mrs. Smith was depressed and socially withdrawn, she was referred for psychiatric consultation. She told the psychiatrist, "I don't like my attitude to the house or to the kids. Something keeps me from doing things for my house or the kids. When I go into the house I suddenly get very tired. I am depressed without cause, confused and always tired."

Mrs. Smith was somewhat blocked in her speech, and spoke in a dull, stilted manner. She said she had had several frightening hallucinations, usually at night. She complained considerably of indecision, even when it came to trifling matters. Of her father, who died when she was sixteen, she said, "I do not remember him very well, except that during my childhood he always seemed to be taking mother away suddenly to go some place."

She described her mother as being a restless, nervous person who was protected and indulged her all her life. According to Mrs. Smith, her next younger brother was the very opposite of herself, "literally a genius." Apparently his arrival on the family scene when she was two had been a traumatic experience for her.

A nervous, frightened youngster, she had few friends in childhood or adolescence, and spent much of her time dreaming. She started going out with her future husband, who was a friend of one of her brothers, in high school. At college, she found it difficult to concentrate on her studies. In the two years following graduation she had ten jobs and was dismissed in almost every instance. The psychiatric diagnosis of Mrs. Smith was that of a schizophrenic reaction.

Mr. Smith described his irritation with his wife's neglect of her household duties. He stated that for the past year he had not been able to get her to take a bath. He was resigned to his wife's lack of desire to be with people and her tendency to sit in a corner by herself when they were in a group. He said their sexual adjustment was "normal," although she had developed an antipathy to sexual relations during one of her pregnancies.

The husband is one of three children. Attached to his brother, he spoke rather contemptuously of his sister, whose intelligence he depreciated. When he was fourteen, his mother committed suicide. He remembered her as an overly clean and meticulous woman, and also said she was antisocial and did not permit him to bring friends into the house. His father was brusque, efficient and not very demonstrative.

As a youngster, Mr. Smith had had several neurotic traits. He was a bed-wetter until the age of nine, and bit his nails throughout his childhood. The brother of his future wife was one of his few friends. Mr. Smith was diagnosed as being a passive character, whose hostility and contempt for women were based on his relationship with his mother and sister. He felt acutely inferior to his brother, and derived considerable satisfaction out of his superiority in competing with his sister, as he later did with his wife.

In the interaction between this wife and husband, it is evident that she had a need to be the child and to be mothered by him. Further, she had a deep masochistic need to be dominated. At the same time, her husband had a need to feel adequate by pointing out his wife's inadequacies and unconsciously keeping her inadequate. Out of his own passive feminine identification, he also had a need to take over his wife's role as mother with the children. As has been noted, the children reacted to the tense interaction with neurotic traits, which only increased parental anxieties and reactions.

In attempting to render a family diagnosis, it is of primary importance to determine the degree of functioning of the marriage or the family, as compared with the degree of neurotic or psychotic manifestation in each member of the family. What is evident in many couples is the need to mold one's partner into the unconsciously expected role of parent. Moreover, the partner is usually unaware of the other's need to re-enact one very essential aspect of the relationship with a parent. The following detail of a case is illustrative.

The husband in this instance subjected his wife to a great deal of humiliation, part of which had to do with his soiled underwear. Ordinarily he threw such garments on his wife's dresser, for her to wash. However, in the course of her treatment she came to realize that it was not realistically necessary for her to stand for this type of behavior. A day came when she was able to pick up the underwear, throw it back on her husband's dresser and say, "You wash it yourself." By doing what she did, she unknowingly completed the re-enactment of an incident in her husband's early life. His mother had died when he was a little boy and his father remarried shortly afterward. Whereupon the child regressed and came home from school every day with fecally soiled undergarments. At home he would take them off and give them to his stepmother to wash. This went on until one day his stepmother, in order to cure him of the habit, forced him to wash his own pants. The repetition compulsion of such infantile memories and anxiety situations in the context of marriage can rarely be overcome without effective treatment.

As for the approach to family diagnosis, the first question psychoanalysis is called upon to answer is concerned with certain dynamics of the marital relationship. What is the conscious and unconscious relationship and interdependence between two marital partners? How do they operate in regard to each other's needs, defenses and instinctual drives? This concept has already become part of our thinking with respect to early childhood and the child's relationship with his mother.

The individual who cares for the child from birth carries out a number of functions that are necessary for the well-being and growth of the child. The reverse situation may also be true. For instance, the parent may unconsciously direct the child into an action which is partly the result of the parent's unconscious need. This acting out by the child of the unconscious needs of the parents is generally considered unwholesome and may have a bad influence later on in the child's life.

This interdependence is illustrated by observations I made on three patients who had been breast fed for a prolonged period, which fact influenced, in part, the oedipal phase and marriage. In one case the boy was permitted access to his mother's breast up to his fourth year. So close was the understanding between child and mother that the words "rush-rush" addressed to the child, as recalled in his analysis, meant "father is coming."

Such infantilization by the mother allows for a stronger fixation on mouth and breast than is helpful in mastering the oedipal conflict. As a result, the homosexual attachment to the father increases, and in this patient's marriage there was strong rivalry with his wife because of his feminine identification. More often this infantilization is observed in the youngest of several siblings. Two out of the three patients on whom I made these observations were the youngest of large families.

Regarding the response of the parents to the infant's or child's needs, we have to determine whether this need-response is wholesome or unwholesome for one or the other, or the family.

Sometimes an understanding of the husband-wife relationship can subsequently be extended to the relationship between parent and child. In one case the patient's behavior *as* a child, which was revealed in his analysis, could later be observed in his behavior *with* his child. The patient was a provocateur—socially, politically and in his professional life. As a child, he would flee from the entangling web of a seductive mother, only to find punishment for his accumulated guilt in a sado-masochistic relationship with his older brother whom he was able to provoke into a white heat of anger, which always meant having to take a beating himself. The patient was unaware of the part he played in provoking his brother until he was well into his analysis. Only then did he describe the situation that existed in his present family.

Here, uncurbed by the patient, the elder of his two boys would torment his wife until she dissolved in tears of helpless rage. As the last step in analyzing this particular constellation, it became possible for the patient to realize that he had been giving unconscious permission to his boy to re-enact the patient's own childhood situation by tormenting his wife, the boy's mother.

The foregoing also illustrates the fact that exposure in childhood to a sado-masochistic relationship does not necessarily mean it will be repeated later on in the marriage itself. There may be several different reasons for this. In this patient's case, he had chosen a wife very much in his own image. The patient's own unconscious needs became manifest only after the birth of his son.

It is precisely because of these considerations that "normalcy" in one or both marital partners is not a necessary prerequisite for a harmonious marriage. An understanding of the compensatory mechanisms is more pertinent in evaluating the constructive or destructive elements in the interaction.

To quote Ackerman, "If one estimates in this manner the assets and liabilities with regard to the mental health of a marital relationship, it begins to be possible to institute a realistic program of therapy. One may then erect goals for the psychotherapy of husband and wife in the context of a clear awareness of the reciprocity of the marital roles, rather than operating psychotherapeutically with an abstract goal of cure of neurosis in one individual partner, while giving insufficient attention to the features of the total marital relationship. One must take into account, too, what is residually healthy in the personality of each partner and the positive factors preserved in the relationship over and above the context of its neurotic component. Adequate diagnosis of the marital relationship makes it possible not only to focus therapy on the damaged areas of functioning but also to reinforce and strengthen the relatively healthy areas of functioning and this tends to fortify the whole family relationship." [1]

Growing understanding of the dynamics of interaction between man and wife is the first step toward inclusion of other members of the family into an integrated plan of treatment. The family includes not only its members in the strict sense— children, parents, in-laws—but "those who lived under the same roof with the paterfamilias; those who form (if I may use the expression) the fireside." [7] Even the maid or the animal pets are psychologically important.

At times the existence of an analogous pathological area in husband and wife may be the cause of a great deal of trouble in the course of treatment as, for instance, when we are dealing with two paranoid persons each of whom uses the marital partner for the projection of his own inner conflict. A *folie à deux* in the couple described below, plus a sado-masochistic relationship, posed a difficult problem in treatment.

A woman married for the second time. She brought into the marriage a son who at the time of observation of the couple is in his middle twenties. The wife and husband had a daughter of twelve. The woman was openly paranoid and accused the husband of infidelity. Her condition became worse when her son married and left the home. She then became openly jealous of her daughter-in-law.

When the wife's condition became so bad that the emotional health of the twelve-year-old girl was threatened, the husband refused to let the girl leave, thereby revealing a similar tie toward the daughter, however covertly, that his wife had openly shown toward her son. The husband's need to use the manifestly psychotic wife as a cover up made it impossible for him either to institutionalize his wife or to let the girl go to a boarding school. Here we are dealing with a sado-masochistic relationship that is capitalized on both by wife and husband for their own respective needs, and where a similar pathological area in both partners is mutually aggravating.

The situation in the following case was further complicated and involved by the role of a housekeeper in the family life. Matilda had for many years been in the employ of the husband's family. When the wife was pregnant, Matilda came to help out. After the patient's baby was delivered, she developed the unreasonable fear that Matilda would take her baby away from her.

Matilda, it turned out, represented different figures for each of these marital partners. The wife, my patient, saw in her the early mother image often found in the unconscious—the baby-snatching witch. For the husband and his family, Matilda played quite another role. His own mother had virtually surrendered him at birth to her housekeeper. Then, the unmarried Matilda, eager at least to have a chance to raise a child, had made the most of the situation and was more of a mother to the patient's husband than his biological mother.

Thus the wife's fear of having her baby snatched from her by Matilda was projected onto the very woman who, in a sense, had taken her husband away from his mother in the first place. For the wife, Matilda remained the personification of everything negative in her own mother. In her analysis, whenever material came up in reference to jealousy, guilt, resentment or hatred toward her own mother, the person who appeared in her dream more often than not was Matilda.

The proper understanding of marriage and family demands, as in the last instance, an intimate knowledge of each relationship. Family interactions are infinitely intricate and complex and require a careful study of the need satisfying economy among all of its members, including in many instances, nurses, maids, and housekeepers.

The following chapters of this volume, dealing with the casework diagnosis and treatment of marital problems, further illuminate the concepts of family diagnosis and dynamics as related to conflicts or to satisfactions in marriage.

References

1. Ackerman, Nathan W., "The Diagnosis of Neurotic Marital Interaction." *Social Casework*, April, 1954.

2. Burlingham, D., *The Psychoanalytic Study of The Child*, Vol. X. New York, International Universities Press, 1955.

3. Dicks, Henry V., *British Journal of Medical Psychology*, XXVI, 1953.

4. Fluegel, J. C., *The Psychoanalytic Study of the Family*. London, Hogarth Press and the International Institute of Psycho-Analysis, 1929.

5. Freud, Sigmund, "Contributions to the Psychology of Love." *Collected Papers*, Vol. IV. London, Hogarth Press, 1934.

6. Jones, Ernest, *The Life and Work of Sigmund Freud*, Vol. II. New York, Basic Books, 1956.

7. Kenyon, Lord, "Judgment." *New Dictionary of Quotations on Historical Principles from Ancient and Modern Sources*. H. L. Mencken, ed. New York, Alfred A. Knopf, Inc., 1942.

Casework Diagnosis of Marital Problems

———— ••• ————

I Psychoanalytic Contributions
SIDNEY L. GREEN, M.D. *

TRADITIONALLY AND JUSTIFIABLY the family casework agency
has been acknowledged to play a principal role among those
community resources dedicated to the preservation and restoration of stable, integrated and healthy family life. The psychosomatic approach in medicine has been paralleled in the modern
concepts of scope and method of practice by which most family
casework agencies have come to be guided in recent years.
These concepts have included an increasing recognition of the
psychological components that contribute to, result from, or are
associated with the emotional and material needs of the people served.

The Division of Family Services of the Community Service
Society has long recognized that the most elaborate of formal
agency services are of relatively slight value unless they are utilized by the casework staff within the proper frame of reference.

---

* Chief Psychiatric Consultant, Division of Family Services, Community
Service Society of New York.

This frame is provided by a thorough understanding of the personality structure of each of the clients served, as well as of the social and somatic factors influencing them. Psychoanalysis has probably contributed more to the achievement of this understanding than any other single approach to the investigation and treatment of human mental functioning. For this reason, for the past fifteen to twenty years the Division of Family Services has continuously made use of psychoanalytically trained psychiatric consultants. During this time it has sought to make use of those psychoanalytic concepts that can be properly and successfully integrated within the substance of casework practice in order to render such practice more adequate and effective. These consultants have contributed from their knowledge and experience through the media of consultations with individual staff members about particular clients and through seminars with groups of caseworkers.

EGO STRUCTURE AND FUNCTION

There are many aspects of psychoanalytic theory and practice which experience has shown can be successfully applied to the casework treatment of marital problems. Certain safeguards, however, must be provided to ensure successful application. Although these safeguards are essential, space limitations permit me to mention only the fact that they consist of continuously available psychiatric consultation and casework supervision of a high caliber in order to ensure the intelligent use of psychoanalytic principles and to guard against the misapplication or distorted conception of such principles.

Perhaps more than anything else in the appraisal of psychological factors, the proper evaluation of ego structure and function offers the caseworker information that is essential to the understanding and management of marital problems. An adequate assessment usually enables the caseworker to make a reliable early estimation of the client's ability to use an interpersonal relationship, and it offers important clues as to the manner in which such a relationship is likely to be used. For these reasons, it also makes possible a decision regarding the more immediate casework treatment objectives and the best methods to be employed.

We have learned to employ certain criteria as an aid in evaluating *the degree of ego maturity and integration.* This is a more precise expression of what the worker is seeking than the ambiguous, variously interpreted terms "weak ego" or "strong ego." Inasmuch as the criteria used are those familiar from the existing psychoanalytic literature, I shall describe them only briefly in order to indicate how the appraisal of the aggregate can provide the caseworker with a most valuable and practical guide to the strengths, weaknesses, and accessibility of his client's personality. Once this knowledge of the individual has been acquired it becomes possible for the caseworker to comprehend more accurately the dynamic pattern of the interpersonal relationship that exists not only between client and spouse, but with other significantly related people as well.

The following outline has been found to provide a practical guide to the most important ego functions. Although these functions are described singly, their apparent isolation from one another is merely superficial, necessitated by the use of the outline method of presentation. It is well known clinically that there is considerable overlapping among these functions and, as a result, they can hardly be distinguished sharply from one another in the healthy, well-integrated ego.

1. *Object Relationships:* On the basis of direct observation and information available from and about the client (and/or spouse) we may ascertain the extent to which the client's libidinal investments either are narcissistically bound to his ego or are made in others in a way that makes warmly emphatic relationships possible. Although there are important exceptions, a serious impairment of this capacity, no matter how subtly concealed by

compensatory character traits, generally promises little likelihood of either an effective casework relationship or an improved marital relationship.

These observations and the information from which they are derived also afford significant clues to the degree of maturity and integrated organization of the client's self-image.

2. *Reality Testing:* This function includes the ability to recognize the actual nature and relative degree of intensity of the feelings manifested in others and one's self, as well as the ability to make the distinction between the actual nature of objects and situations on the one hand and fantasies on the other.

3. *Judgment:* This function is closely related to reality testing. It involves the ability to retain the correct impressions and arrive at logical conclusions from past experience and to integrate these impressions and conclusions with those gained from current situations. When such retention and integration are adequate, we may expect the individual to be able to arrive at an appropriate plan with regard to present situations and future related ones. When judgment is adequate, there will also be the ability to evaluate reliably the extent to which the participation or assistance of other people should be sought or avoided. This aspect of judgment is closely related to the kind of object relationships which are established by the individual being studied.

4. *Motility Patterns:* In order to evaluate this function we note the adequacy, appropriateness, and range of the kind and amount of activity (or inactivity) with which the individual meets life situations. This appraisal is made in regard to responses that the individual makes to the subjective as well as to the external stimuli with which he must cope.

5. *Tolerance for Frustration:* Two factors must be noted in order to assess this function properly; they are of equal importance. The first is the ability to postpone the need for gratification of impulses. The second is the extent to which the accustomed freedom of general ego functioning remains unimpaired when some degree of postponement is accepted. The overlapping

relationships among this function, judgment, and motility patterns should be noted.

6. *Affectivity:* In this connection we note range, adequacy, appropriateness, and lability of affect which the client is able to mobilize. In the presence of any indication of a mood disturbance we must look for signs of pervasive moods, since these offer valuable clues to mild or early depressive or manic states.

7. *Defense Mechanisms:* As the contact with the client develops, the caseworker is constantly alert for repetitive, patterned behavior responses, since they are usually indicative of a limitation of ego adaptability which is due to the *unconscious* restrictive influence of defense mechanisms. When trying to identify the mechanisms operating, the worker must be able to make a clear distinction between defensive *behavior patterns* and the defense mechanisms that may be responsible for them.

8. *Basic Intellective Capacities:* In this category we attempt to determine the potential *inherent* capacity of the client to make profitable use of his endowments and experiences. The administration of a battery of psychological tests may be necessary in order to be able to make this determination accurately. The findings on the formal psychological examination are often quite helpful for determining how much of his intellective capacity an individual is actually using.

If the usefulness of the foregoing criteria has been fully exploited, the caseworker is now in a position to begin to estimate accurately the degree of maturity achieved by the client and how well he may be expected to cope with the stresses of his everyday living experiences.

However, psychoanalytic metapsychology has shown that there is more that must be known before we are able to make a tangible contribution to the needs of emotionally troubled people —in this instance, married couples.

We are well aware of the fact that it is the skillful exploitation of the positive transference—in the broadest sense of the term—which enables one person to influence the way of living of another person by means of discussion techniques. Such discussion techniques seek to ensure the use of mature logic, intellect, experience and judgment as the dominant forces that determine the client's ways of living in relation to others. In casework literature this broad concept of the positive transference is included in the concept of the positive relationship. The casework method of treatment does not aim toward helping the client to understand his transference or toward the lifting of repression from hitherto unconscious instinctual needs. It is an established fact, however, that the most effective casework treatment is conducted by those who are most sensitively aware of the pattern of the client's transference and the nature of his principal unconscious conflicts. For example, there are some instances in which the sex of the worker may make the difference between successful and unsuccessful casework treatment because of the client's unconscious needs. In other instances, the outcome of casework treatment in marital problems may be determined by whether there is an assignment of the same worker or different workers to each spouse. The proper decision regarding assignment can be made only when, along with other pertinent considerations, the unconscious needs and conflicts of the spouses are adequately known and understood by the caseworker.

The skillful caseworker must have an awareness of the balance existing between the libidinal and aggressive drives in the principal intrapsychic conflicts that have helped to shape the client's personality; of the extent to which the client's ego can or cannot tolerate stimuli that are reminders of these drives. From his ego appraisal, combined with other data from the anamnesis, the caseworker will usually have acquired enough knowledge to enable him to gauge the extent to which his client's personality handicaps have resulted from inherently poor ego integration and/or from neurotic methods of adaptation.

It is often necessary for the caseworker to distinguish between pre-oedipal and oedipal elements in the behavior of the client. When the client's relationships and goals are determined very largely by pre-oedipal needs and conflicts, the worker must usually be prepared to take a more active and assertive role with regard to guidance and counseling. Such clients generally have more poorly integrated ego structures and much greater dependency needs than do those whose patterns of adaptation are based almost entirely on the concepts, attitudes and goals that are characteristic of the later oedipal phase of personality development. The services of the psychiatric consultant have been valuable in assisting the caseworker to make this distinction.

There are also occasions when the caseworker is greatly assisted by the recognition of predominantly anal or oral patterns of adaptation in his client. However, important though it may be to the casework treatment plan, such differential recognition can sometimes be quite difficult to achieve. Here again our psychiatric consultants have been of help to our casework staff.

I have dwelt at greatest length upon ego appraisal because I believe that this is a particularly essential skill for all casework treatment and the one that, fortunately, is most easily taught within the frame of reference of casework agency experience. Successful treatment requires that the caseworker possess a fairly thorough knowledge of the significant conflicts in the client's conscious and *preconscious* psychic life. In this connection, I wish to emphasize the occasionally overlooked fact that the goal of casework treatment is not to achieve insight (on the part of the client) into the nature of the transference, to lift repression from unconscious instinctual drives, or to direct the client in a search for the genetic aspects of his unconscious conflicts. In the following section there is a discussion of the actual nature of casework treatment in relation to marital problems and examples of the contributions made to it by applied psychoanalysis.

As a result of the dynamic understanding afforded by the psychoanalytic approach to human personality problems, we have been able to classify, in certain broad groups, the most commonly encountered marital problems. These groups will be identified and described along with representative examples of each in the subsequent section. At this point, however, I wish to call attention to those particular psychoanalytic concepts that have been useful in establishing the various categories of marital difficulties.

A reasonable understanding of the principal unconscious needs and motivations of the client is essential for successful treatment of his marital problems. It is only in the light of this knowledge—as related to other pertinent factors—that his current life situation can be properly comprehended. Such comprehension also requires a reliable estimation of the degree of ego maturity and integration achieved by each of the partners. Thus we are able to recognize the dynamics of the kind of equilibrium they have achieved. The existing or potential equilibrium may be further determined by other psychological factors such as mutual dependency of the spouses' egos and influential unconscious fantasies, and also by the pressures of environmental stresses over which they have no control.

FUNCTIONS OF THE PSYCHIATRIC CONSULTANT

I have commented on the fact that the Division of Family Services has made use of psychoanalytically trained psychiatrists as consultants for many years. There is an established conviction in the Division that psychoanalysis has an essential contribution to make to casework practice and that the service of psychoanalytically oriented psychiatric consultants is one effective way of providing it. With the passage of time it has become recognized that such consultations are most effective when the psychiatrists are themselves thoroughly familiar with the nature of casework treatment and with the specific structure and resources of the Community Service Society and its Division of Family Services. Under the system currently in operation, a group of psychoanalytically trained psychiatric consultants regularly gives a specified block of time each week in the district offices and in the Youth Bureau. It must be emphasized that they do not function as treatment supervisors. Through the services of these consultants we try to ensure the sound application of psychoanalytic psychiatry to casework treatment and to guard against the danger of the equivalent of "wild psychoanalysis." Consultation on cases involving marital problems is one of their regular functions.

The Division's psychiatric consultation program assumes that a competent caseworker can, in most instances, be relied upon to present an accurate description of a client and his circumstances. It is for this reason that psychiatric diagnostic interviews with clients are not routine matters. Furthermore, psychiatric consultations are not the rule in every case. As a matter of fact, there are not enough psychiatrists available even if this were desirable. In practice, we have found that there can be considerable carry-over to cases with similar problems when psychiatric consultation is provided for representative cases. Such consultations provide both consultant and staff with an opportunity to check on the accuracy and extent of the application of psychoanalytic psychiatry to the casework treatment of marital problems as well as to the other problems with which a family casework agency is prepared to deal.

An additional important contribution of the psychiatric consultant must not be overlooked. As a physician, he also helps with the recognition of somatic problems and the part they may play in relation to the physical and mental health of the clients receiving service in the Division.

It has been my purpose, in this presentation, to define the psychoanalytic contributions to the casework treatment of marital problems, and to indicate how one agency has ensured constant access to those contributions. It is my belief that these psychoanalytic contributions have significantly increased the effectiveness of the casework treatment of marital problems offered by the Division.

II   *Application of Psychoanalytic Concepts*
JEANETTE REGENSBURG, Ph.D.*

BEFORE I ATTEMPT to illustrate the application of the psychoanalytic concepts described in the foregoing section, it is necessary to describe briefly the conceptual framework of social casework treatment within which they are utilized in the Division of Family Services, Community Service Society.

As a family agency, the Community Service Society is charged with giving direct service that will improve the functioning of the family unit as well as the functioning of the individual members of the family. Thus the caseworker is committed to understanding and evaluating the total family situation, whatever the nature of the presenting problem, in accordance with the significant psychological, cultural-socio-economic and somatic factors inherent in that situation. The range and intensity of the study process that leads to such evaluation are matters of professional judgment, case by case, but it is understood that the selection of treatment aim, goals, and techniques[1] must be based on adequate and appropriate case data, accurately evaluated.

For the purposes of this study, emphasis will be placed upon one segment of the total family situation, the marital relationship, although a complete discussion of that aspect would include consideration of parent-child relations as well. Furthermore, this presentation will be concerned chiefly with problematic marriage relations that come to the attention of the family agency. "Relations" implies at once that the caseworker's task of evaluating case data does not end with the assessment of the ego maturity and integration of one or both partners in the marriage, but must proceed to an assessment of the relation between the

* Casework Associate, Community Service Society of New York.

spouses. The evaluation of the marital relationship requires both longitudinal and cross-sectional history; and it cannot be over-emphasized that courtship history is an integral part of the longitudinal marital history. In addition, such data as which spouse makes the application for casework treatment, the motivations of one or both spouses for coming to the agency, or of one spouse for avoiding contact, and the precipitating factors in the application for service, are also essential to the assessment of the marital relationship.[2]

More specifically, what is meant by an assessment of this relationship? What the caseworker comes to know about an individual spouse gives him some information about that partner's consciously and unconsciously expressed needs. In order to assess the relationship between the partners, however, the case-worker seeks to know still other things—in what ways each partner gratifies or does not gratify the other's needs; whether the total current picture of gratification shows balance or imbalance; whether that balance or imbalance springs from predominantly healthy or unhealthy needs and gratifications; how the current picture compares with the past, as well as when and why a change in balance occurred. This kind of assessment thus takes the descriptive data of individual behavior and adaptive patterns one step further.

What is meant by balance and imbalance in gratification is illustrated by one young couple, married six years and with two children. The husband is several notches above his wife in educational experience and for the first years of the marriage she was happy to deprive herself for his advancement. She was thrifty in household management, made her own and the children's clothes, and did not resent her slightly dowdy appearance. All the money she saved went into her husband's well-stocked wardrobe and a constant exchange of one automobile for another newer and better one—for the avowed purpose of improving his business opportunities. Quite gradually it dawned on her that their financial and living conditions had remained

246 «   Neurotic Interaction in Marriage

static; that her husband was, for his part, enjoying the fruits of
her thrift for his personal gratification, not to increase the standards and pleasures of the family group. At this point of her
awareness that gratification was not mutual but all on her husband's side, manifest family tension due to the imbalance began. It was well advanced by the time the wife asked for casework treatment to help save their marriage. Even with these
superficial data, the current imbalance can be recognized as potentially healthy rather than unhealthy, although one cannot
make a prognostication of the outcome of treatment without
many additional data about each of the partners.

Although the precipitating factor in applications is frequently
"the big fight"—over money, a child, an in-law—it just as frequently takes the form, as in this family, of one or both spouses
being "fed up" after a long period of accumulated anxieties, real
deprivations, or fancied grievances. There is not always a change
from balance to imbalance as with the couple described above;
a change in the degree of imbalance may result which makes a
hitherto tolerable situation intolerable for one or both partners.
Clearly, the caseworker cannot make an assessment of marriage
relations against an absolute concept of what is or is not gratifying, healthy, or tolerable in marriage; the clients provide their
own standards. It is our hope, however, that many clients will
be motivated toward the modification of their needs and gratifications in the direction of greater health and more satisfactory
family living.

To have assessed a problem in marital relations does not inevitably lead to treatment of it in the family agency. There are
instances in which the method of choice is psychiatric treatment
of one or both partners rather than social casework treatment;
there are instances in which the focus of social casework treatment should not be on the marital problem—perhaps never, perhaps not until much later—but on related problems. In each
family the choice to continue or to dissolve the marriage rests
with the clients. In his work with the respective partners the

caseworker will help the individual to utilize his healthiest drives and aims and these may lead to dissolution of the union. But whether the choice is for continuance or for dissolution of the marriage, changes made in the relationship by one or both partners must be consonant with the individuals' desires, needs and capacities rather than with an abstract notion the caseworker may have of a universal concept of personal maturity and of a mature relationship.

Between the caseworker's first evaluation of the situation and the clients' attainment of the ultimate or end goal lies a series of intermediate goals related strictly to the individual family and marital situation and to the characteristics of the members that comprise the family. These intermediate goals might be called technical goals in so far as the caseworker, with the professional understanding, planning, timing, and selection of aim and techniques, bears a large part of the responsibility for helping his clients move step by step from "here to there"—from where they entered treatment to where they are ready to terminate contact.

TECHNICAL TASKS OF THE CASEWORKER

At this point it may be well to look at the technical tasks which, although not unique to the casework treatment of marital problems, are highly important and difficult to perform. Since there is always some imbalance or threat of imbalance in a marriage relationship problem, the worker has to envisage what changes can be effected by each client to afford a more tolerable and mutually gratifying relationship. Some changes must take place whether the end goal is continuance or dissolution of the union, since even to arrive at the decision to do one or the other requires a concentrated period of casework treatment in many instances. These are the changes that can be thought of as intermediate or technical goals, for example, in one union there needs to be a shift within the axis of dependency-independency; in another the shift required is within the axis of giving-taking; in still another, within the axis of submission-dominance, and so

on. How to help the clients achieve such shifts depends very
largely on the caseworker's understanding and use of the concepts Dr. Green has presented: the assessment of each partner's
ego maturity and integration; the assessment of the conflict situation between libidinal and aggressive drives in each partner; recognition of the pre-oedipal and oedipal elements in each partner's behavior; determination of what is accessible to treatment
in each partner's conscious and preconscious psychic life and of
what is quite evident to the caseworker but unconscious to the
client; evaluation of environmental stresses, their amenability to
change, and their influence on the client's personality and functioning.

On the basis of these assessments, it lies within the professional responsibility of the caseworker to decide whether one or
both partners must become primary clients; for example, must
they both become equally involved in treatment? Is it indicated
that one must become involved before the other starts; that only
one is to be involved, the other not; that only one is to be in
the focus of treatment and the other on the periphery? How frequently shall each be seen?

Another set of tasks surrounds the choice of worker. Here we
are faced with the choice of worker's sex for one or both partners, as well as with the difficult question of whether there
should be a different caseworker for each spouse or the same
caseworker for both. In general, the choices will depend largely
upon each client's capacity for object relationship and for reality
testing, his affectivity, and of course upon the nature of the relation between the marriage partners.

In the actual handling of the interview situation through the
medium of client-worker relationship, there are a few tasks that
have special significance in the treatment of marital problems. The mastery of these tasks is greatly facilitated by the caseworker's knowledge of and ability to utilize the psychoanalytic
concepts Dr. Green describes. The first task is the caseworker's
obligation to be aware of and in control of transference and

counter-transference situations: to distinguish between those elements of client-worker relationship which are based on reality experiences, past and present, and those based neurotically on the past; to avoid under- and over-identification with either partner in the marriage; to guard against psychologically seductive relations; and to develop empathy with both partners, whether or not both are clients.

A second task in the treatment process—whether each partner is a client or not, whether there are one or two workers—is to help the two partners to communicate with each other; they must be helped and encouraged to discuss with each other appropriate and essential matters. As long as a union exists there can be no constructive resolution of the marriage problem without interchange between the spouses. The technically unskilled caseworker, or the caseworker lost in a transference-counter-transference situation, is apt to be used by the clients as umpire, and go-between, or as a source of personal gratification, and is headed for an ineffective if not a frankly disastrous outcome of treatment.

The third task of special significance occurs only when both partners are in contact with the agency and confronts us whether the spouses are seen by the same or different caseworkers. This task is to use, with each partner, material obtained from the other partner, in such a way as to help both spouses attain their goals without betraying the confidence of either one. There is no intention here of ruling out the caseworker's direct and overt use of a client's communications on occasions when it is appropriate and agreed upon by client and worker. There are many occasions, however, when such direct and overt use is undesirable, destructive to both the marriage relationship and the client-worker relationship, or unnecessarily painful to one or both partners. It is in these instances that the caseworker is called upon to make indirect yet effective use of what he knows. He avoids becoming a prosecutor, a protagonist, or a defender, but at the same time can help both spouses control their neurotic demands upon each other and afford each other increasing gratification.

An informal review of situations in which the central problem focused on the marital relationship resulted in the formulation of several categories of marital relationships. These categories, along with appropriate illustrations from case records, are offered for consideration and testing with full awareness that the range of possibilities has barely been tapped and that the validity of the approach must also be further confirmed.

1. *Where Tensions Are Gratifying:* In this category are found marriage relationships in which tensions are mutually gratifying, consciously or unconsciously or both. In these situations there seems to be a mutual dependency growing out of the support each spouse obtains from the other for his unhealthy needs. The tension itself gratifies a neurotic need. In these marriages there has often been an unhealthy balance at some earlier period which, at the time of application to the agency, is more overtly disturbed, impairing the functioning of the family unit and its members. The usual pattern in this kind of relationship seems to be one of living-in-conflict and the partners are ordinarily people whose ego maturity and integration are markedly inadequate, but who desperately need each other. The casework goal is consequently to restore the old pathological balance and thus effect some improvement in each individual's and the family's functioning. There can be no resolution of underlying conflict or change in the basic neurotic, defensive, adaptive patterns; on the contrary, reinforcement of useful defense mechanisms should be effected. The following account is an illustration of this type of marital situation.

A childless couple had been married for ten years when the wife came to the family agency, asking for help. Because the husband was in the service during the first few years of the marriage, they had maintained a home together for only the last five years. The wife explained that her request for help was due to her husband's fear of other people, his feeling inferior to her, his rages in which he struck her, and his subsequent remorse and exhaustion. The caseworker, a woman, noted that the wife's tone was one of a mother describing a naughty but rather amusing child; she laughed about her husband's rages, thought she should be more patient with him and not dominate him as much as she did.

The wife continued regular and frequent contact with this caseworker while the husband, who came in after considerable pressure from his wife, maintained a less frequent contact with another woman caseworker (he himself had requested that he see a woman). It was obviously agreeable to each spouse that there be two workers and this was considered professionally sound in the light of the clients' general immaturity and weak capacity for object relationships. The slight identification of one with the other would have made the sharing of one worker a potentially destructive experience. Also, the wife's possessive attitude toward her husband was an omen of intolerable competitiveness if they shared the same worker.

It was learned that the wife worked while her husband attended school on his veterans' benefits. He did all the shopping, handling of money and paying of bills. She disliked housework, but did the cleaning and cooking under protest. He had many friends and led an active social life apart from his wife. She had almost completely dropped former friends and interests.

Separations for a day or two at a time were frequent, the wife walking out on him on these occasions. Both were unhappy and depressed when separated, but usually it was he who begged her to return. In discussing the marriage, the wife told her caseworker she *had* to be the dominating person and did not want it otherwise. She knew before they married that he evaded responsibility and got away with things; she thought of him, when she saw him in his own home, as a mischievous child who needed to be controlled. Once when the husband discussed his father's and wife's bossiness he slipped and called her "my mother." In their physical fights the wife was the one who would punch while he scratched and slapped. He would get exhausted before she did, and when she had regained control they became tender and solicitous toward each other.

At the time of application, the state of affairs was such that the wife was frightened by her husband's increasing loss of control and the strength of her rages in response to his. The hitherto tolerable sado-masochistic situation was so out of control that it was no longer gratifying.

The outstanding features of this relationship were the wife's need for gratification of her impulses to control and belittle her husband, and his need to be disciplined and humiliated by her. The psychiatric consultant confirmed this evaluation of the relationship as well as the caseworker's recognition of the latent homosexual components in each spouse's personality. It was these components that contributed largely to the balance, however unhealthy, which had obtained for a period of years in the marriage. The imbalance that brought the wife to the agency had been precipitated by the excess of aggression each partner was experiencing and the reactive responses of each to the other's increased aggression. The wife was the key person in that her anxiety motivated the application. Reduction of her aggressive behavior would automatically reduce his and thereby, hopefully, restore the status quo.

On these premises, casework treatment was aimed at strengthening the wife's defenses by redirecting some of her controlling drives away from her husband onto her job, onto resuming social contacts she had dropped quite completely for the last several years, and onto some cultural, avocational interests she had had in the past. Development of self-awareness was not indicated since it was evident that both these partners had a very tenuous hold on reality, and were likely to break down either if they lost each other's support or if there were any breach in their already weak defenses against their anxieties. Furthermore, deliberate attempts to increase the self-awareness of either partner would have strengthened neurotic transference elements in the client-worker relationship which called, instead, for as firm a basis in reality as it was possible to achieve. Treatment of the husband, too, was directed toward improvement in his functioning in his vocational courses and toward strengthening his determination to seek employment at the earliest practical opportunity.

Over a period of about a year's contact the intensity and frequency of the rages and physical fights were reduced, and there was some progress toward better functioning in the clients' social life and vocational settings. The interdependence for mutual gratification of their essentially unhealthy needs was maintained without any change in their habitual adaptive patterns.

2. *Where Tensions Are Intolerable:* There are marriage relationships in which tensions are ungratifying and intolerable, and in which change is desired by one or both partners. In these situations each partner may or may not be capable of meeting the mature aspirations of the other. The inherent capacities of each spouse to afford the amount of gratification required by the other must be discovered and evaluated so that reasonable goals can be determined. Treatment may result either in a more satisfactory continuance of the union or in a relatively healthy dissolution of the marriage, depending on the aforementioned capacities to afford mutual gratification. The example that follows illustrates a situation in which change was desired and achieved by both partners, with continuance of the union.

This family consists of mother, father, and one preschool child, a boy. When first applying for help, the parents were concerned about the child's aggressiveness, which they could not handle. Rather quickly both mother and father recognized that the situation between them was paramount and for about a year their use of the caseworker's services was directed toward improving their own relationship. The marriage was based on mutual love and consideration which encompassed love for their child and a desire to create an harmonious family unit.

The partners' expressed dissatisfactions were these: the wife was irritated by her husband's excessive investment of time and interest in his job, by his allowing an inconsiderate, unreasonable employer to exploit him, by his exaggerated ideals of perfection for himself, and by his tendency to extreme fluctuations in his relationships with other people, especially his siblings. The husband's description of his dissatisfaction in the marriage indicated chiefly a lack of free communication between himself and his wife, a tendency on her part to leave him to his own devices and to refrain from giving an opinion on matters that affected both of them. He was also disturbed by his excessive feeling of responsibility on the job, which he recognized as unreasonable but could not modify, and by his general tendency to demand perfection of himself.

In essence, the underlying problem which, at the time of application, was interfering with more complete mutual gratification can be described in this way: the wife, consciously determined not to become like her mother, who controlled and directed her father, was in great fear and restraint of her own aggression, even the most ordinary expressions of difference, opinion and suggestion. Her very self-control deprived her husband of support, advice, and shared responsibility which would have gratified his strong underlying need for some dependence and some release from his painful efforts to be a perfect, utterly independent human being—an ideal he created chiefly out of his own father's demands on him in childhood and early youth.

Each partner was an adequate person, with sufficient emotional maturity to work with the same caseworker and with each other toward the same goal. In this case, too, the psychiatric consultant contributed toward the evaluation of the individual personalities and of the marriage relation.

With these people's inner strengths, some change in basic defensive adaptive pattern was possible of achievement and the technique of clarification was therefore dominant in the treatment process. For each partner, after recognizing the inappropriateness of some of his behavior to current reality, the next step was to recognize the relation between his significant remembered childhood experiences and his current behavior toward his spouse; and next, to utilize his knowledge to modify his behavior and both give and receive increasing gratification. The wife became able to express an opinion, give some sound advice, and voice an objection, as any marriage partner should feel free to do. The husband, after some attempts to make his job operable, finally changed it to one much more satisfactory in hours, salary and general atmosphere. During the course of treatment there was a notable increase in communication between husband and wife not only about the usual responsibilities of family life but about themselves as acting and reacting human beings.

It is worth noting that, although the marital relationship gradually moved into equilibrium, the child did not automatically cease to be a problem to his parents. They had still to learn how to make comparable changes in their relations to him. In the third stage of treatment the mother learned that, influenced by her own fears, she had held the boy back from natural expressions of his aggressiveness and from attempts to master real situations. In direct contradiction, the father's chronic drive for perfection was pushing the boy to most unreasonable standards of mastery and achievement. Neither parent was, in effect, treating the son as a child. Both are now well on the way toward releasing him from their unreconcilable demands.

The following is an example of a family in which change was desired by one spouse, resulting in a healthy dissolution of the marriage.

There were originally a mother, father, and four children of school age in this family, and the home was maintained on a public assistance grant. The father, who had been ill, was considered by physicians to be able to return to work, but he refused. The mother was beset with worries about her husband's unemployment, the difficulties in managing on a marginal income, behavior problems manifested by the children and her own impaired health. What emerged at first was the mother's self-image as a disparaged woman who doubted her capacity to function as wife and mother. The father was revealed as a severely disturbed man whose overt behavior was punitive and physically dangerous to his wife and the children. The psychiatric consultant offered a tentative diagnosis of paranoid schizophrenia for the father. The father had, by plan, only intermittent contact with the agency.

This woman did obtain unconscious gratification from suffering, which was all too liberally supplied by her mentally ill husband. As she suffered and felt oppressed and depressed she drew the children to her more closely as her only source of conscious pleasure; and they in turn were drawn further and further from normal and healthy activities.

Fortunately, however, the client's conscious dissatisfactions were healthy and served as leverage for bettering her condition. She shifted in the course of casework treatment from bitter complaints about her husband's behavior to a decision that if anything were to be improved *she* would have to make the changes. Her growing self-awareness of her role in the family situation was encouraged not only by the caseworker but also by a close relative who told her she was "always wanting to stay in a bad situation." As the client related more rationally to the fact of her husband's illness and its consequences, she began to use the caseworker's guidance in budgetary and household management, accepted the offer of activity group therapy for the two older children, and sought healthier recreational outlets for the family. In general, she found herself able to create a situation which had in it more pleasure and less pain.

Eventually, she made the decision to obtain legal separation and took the necessary steps, including the difficult ones of explaining her decision to the children and to her husband. Her progress was the result of strengthening her healthiest impulses and channeling them into more highly socialized activities. The caseworker sanctioned pleasure and comfort, redirected the client's unhealthy protective attitudes toward the children into healthier ways of "giving" to them, recognized and encouraged her potentialities as a responsible and able woman. Not the least of the services was the caseworker's recognition of socio-economic pressures and her help in removing and reducing them.

The children's expanded interests and activities gave them some healthier images of men and father-persons and opportunities to make healthier identifications with adults of both sexes.

The gist of the improved functioning of the family members and of the family unit after the union was dissolved can be summed up by two remarks the mother made in a planned follow-up interview about a year after termination of contact. One was that she had gained thirty pounds since the day of her application to the agency; the other comment was, "The house used to be so quiet people never knew there were children. Now there is noise and there is no doubt about it. I like it that way."

3. *Where Mutual Gratification Is Disturbed By Crisis:* The marriage relationships in this category are those in which there is mutual gratification and a consistent balance—either healthy or unhealthy—but in which a crisis has occurred, often due to unreasonably severe external pressures. At times this pressure results from an intrapsychic conflict in one spouse whose otherwise successful defenses have weakened. In this category the goal of treatment is to resolve the problem situation, leaving the relationship itself intact. In other words the attention of caseworker and client is centered on the illness, financial problem, and so on, or, if appropriate, on restoring the strength of weakened defense mechanisms. Thus the threat of imbalance in the

relationship is removed and each individual can be restored to his own maximum functioning. The following illustration is one of an emerging intrapsychic conflict that called for a stronger defense.

In this situation, for which there was no psychiatric consultation, the wife solved her immediate problem to her own satisfaction in one interview. She stated her purpose in coming to the agency in very direct terms: should they or should they not have a second child? In the mothers' group discussions that she had been attending it had rather suddenly occurred to her that they ought to have another child. All the other mothers had two or more children and she was getting on in years. Her husband did not agree with her at first but she "brought him around." What motivated her to come to the agency was the degree of discomfort she now felt after he assented to her wish.

In this one interview she explored various aspects of their placid life together, rather surprised to discover that she had made most of the major decisions, that she paid little attention to her husband's business and avocational interests but did, as he said, devote an extraordinary amount of time to their child. She summed up what she had clarified for herself by recognizing that her discomfort sprang less from the fact that she was not herself so anxious for another child as she had supposed, than from guilt that she had coerced her husband about this issue and had tended in the past to dominate and at the same time to neglect him. She made her decision to tell her husband that she had changed her mind about having a second child, remarking with some humor that she would feel very silly but it was the thing to do. She left with the understanding that she would return if she had further need for the agency's services.

One might speculate that in this instance the client's usually successful defenses against anxieties related to being a woman had been weakened in the group discussion. In the course of her conversation with the caseworker she seemed to be reinforcing her controls over those anxieties by a kind of displacement to another closely related area of her life, more easily faced and handled, more acceptable to her ego and image of herself as a woman. In other words, she need not have the baby neither she nor her husband really wanted. Instead she would reduce her guilt by modifying her controlling behavior toward her husband and the child they had, thus removing the recent threat to balance in the marital relationship.

4. *Where There Is Mutually Gratifying Attachment:* This group of marriage relationships is one in which two people have lived for a time in physical proximity without manifest interpersonal relationship but whose comfort is disturbed by a change in the family situation. These unions seem characterized by a lack of identification of one partner with the other, each partner apt to be detached, remote, and narcissistic. The threat to this unhealthy kind of balance often appears to be the wife's pregnancy or the birth of a child, since such an event is a severe disturbance to each partner's narcissistic gratifications. It is nip and tuck whether these unions continue at all. One has to remember, though, that extreme narcissism is itself the expression of a need and attempt at fulfillment and that many clients so characterized can respond, a little and slowly, to the caseworker's empathy, outgoingness, and practical offers of help.

One such married couple was described by the caseworker as resembling "two students who have taken an apartment together." To all intents and purposes this young man and woman were satisfied to have a comfortable living arrangement for which they shared various household and financial responsibilities, but in which they did not seek nor feel the need for mutual gratification of emotional needs. When the woman became pregnant, her husband threatened to leave, and it was her fear of the financial strain—since she would have to stop work for a while and did not in any case wish to support herself and a child—that was the conscious precipitating factor in the wife's application.

Finally, there is one kind of marital situation, not in itself easily assessable, which justifies the caseworker's attention. This is the marriage relationship in which the union has been dissolved by physical separation of the spouses but in which the psychic bonds have not been severed by one or both of the partners. In this kind of broken union one of the partners has an active place in the mental and social life of the other partner and of the children, if they exist. Usually the wife and children are known to the family agency in these situations, and one cannot always know whether the fantasy-union exists also in the life of the separated husband and father. In these unions—which really are not broken—the original tensions live on and are manifested in the behavior and verbalizations of wife and children. The fantasies, the psychological needs unmet in reality, are recognizable to the caseworker even when not overtly expressed.

In some of these cases the physical separation, once it is made, seems to be final. But we are all familiar with the family that chronically separates and unites, time and again. Further exploration might indicate that some of these relationships belong rightly to the first category proposed. Living-in-conflict is the pattern, becomes after a time intolerable, is again needed and sought after.

It would, therefore, seem obligatory for the caseworker to understand and assess the marriage relationship that existed prior to separation. Often, for example, the caseworker learns that the wife uses the children to appeal to the father, or to demand help from him, when there is a financial crisis, illness, or other event that can be exploited in the service of the wife's ungratified needs.

Summary

In recent years, caseworkers and psychiatrists engaged in their respective roles have become increasingly interested in sharpening their ability to evaluate the family unit. This desire for a system of family diagnosis, as we often call it, is not an aim in itself but a necessary base for improving treatment techniques and outcomes. This chapter suggests one way of classifying problems in marriage relationships which takes into account the dynamics of intra- and interpersonal psychic action and reaction. It is hoped that the general approach is worth experimenting with as we continue in our struggle to go beyond the conventional pattern of problem classification.

References

1. For definitions of social casework terms, see *Scope and Methods of the Family Service Agency,* Report of the Committee on Methods and Scope. New York, Family Service Association of America, 1953.

2. Ackerman, Nathan W., "The Diagnosis of Neurotic Marital Interaction." *Social Casework,* XXXV, No. 4, 139-47, 1954.

General References

Berkowitz, Sidney J., "An Approach to the Treatment of Marital Discord." *Social Casework,* XXIX, No. 9, 355-61, 1948.

Josselyn, Irene M., "The Family as a Psychological Unit." *Social Casework,* XXXIV, No. 8, 336-43, 1953.

Hollis, Florence, *Women in Marital Conflict,* Family Service Association of America, New York, 1949.

Weiss, Edoardo, *Principles of Psychodynamics,* New York, Grune & Stratton, 1950.

Casework Treatment of Marital Problems

1 Evolution of Treatment Methods

FRANCES LEVINSON BEATMAN*

A HISTORICAL PERSPECTIVE on the development of casework techniques may be useful in understanding the more recent trends in family agency practice. Social work, in its pre-professional era, was first undertaken by community leaders who felt impelled to assist the less fortunate but who had no special training for the task. Even in those early days, however, the seriousness of problems stemming from conflicted marital relationships was recognized, and efforts were made to cope with them.

Over the years, as communities became more aware of the various sources of infection that attack the integrity of marriage and the family, a variety of institutions arose to meet different aspects of the problem. Each successive period created new services, aimed at salvaging or correcting the consequences of breakdowns in family life. Gradually, these methods changed from

* Associate Executive Director, Jewish Family Service, New York City.

services rendered by lay people to the present approaches which are based on professional standards of training and service.

The earliest attempts, having been influenced mainly by sociological concepts, moral concerns and ethical principles, sought to support troubled individuals by manipulating their environments. Since 1920, however, the underlying philosophy of treatment has been to integrate knowledge of social forces and manipulation of environment with the dynamic principles of personality learned from psychoanalysis.

Traditionally, the family agency has acted as guardian and counselor for families who found themselves unable to manage their own affairs for a variety of reasons—economic and vocational inadequacies, physical disabilities, personality difficulties, emotional immaturity and various combinations of these problems. Also, the family agency was considered the first source of aid in times of war, depression and other major catastrophes.

In the course of performing these functions, family agency workers have had almost unlimited opportunities to observe the marital relationship in action, in healthy as well as in troubled marriages. We have seen marriages under stress because the partners were unable to meet the demands of a complicated adult society, and marriages which somehow managed to survive crises and economic hardships.

Sometimes, previously satisfactory marriages have broken under the strain of economic deprivation or changes in the cultural and social setting. By contrast, other unions seemed to take on their first real seasoning and maturity in a critical situation. Still other marriages could not survive the inability of one or both partners to grow with the marriage, while some couples adjusted well to continual changes in the family constellation and environment.

During the depression of the 1930's our attention was focused on understanding the needs of the individual beyond economic factors. While rendering practical assistance, we observed that in many cases the economic crisis was used as a channel for the expression of latent unresolved fears and anxieties. This realization led to more complete consideration of the balancing forces in interpersonal relationships—the attitudes of family members and their interaction within the total family constellation.

As time went on, it became increasingly clear that our casework treatment plan for dealing with marital troubles had to be based on a diagnostic understanding of the marriage, as well as on our diagnostic appreciation of the individual marital partners. In other words, we had to examine the emotional needs of the couple involved, the extent to which they were being fulfilled by the marriage, and how these needs could be met by other life roles—that of parent, worker, community activity and the like. Further, it was important to estimate the extent to which the personality needs of one partner could be fulfilled by the other and the personality weaknesses offset, and whether the marriage was being overburdened by projections from previous unresolved relationships.

Thus the seeds were sown for a three-dimensional picture in casework treatment, integrating the following:

1. Social factors and their impact on personal and family functioning.

2. The personality structure of the individual.

3. Interaction, either as a resource or as an impediment to family stability.

In many situations the casework approach was to support the existing pattern of relationship. In others the goal was a modification of behavior patterns based on an understanding of the personalities, the latent conflicts mobilized by the situation and the effects of weakened defenses.

In the decade after the depression, marital counseling became an important function of the family agency. With financial relief taken over by civic and governmental agencies, the family agency was free to work with psycho-social problems. The growing sophistication of the community made it possible for many people to seek help in meeting their family responsibilities more adequately. At the same time, the family agency made itself more available to all economic groups by instituting payment for service by those who could afford to pay.

Today, many applicants who come to the agency are not under undue social or economic pressure. Some come in the early stages of their marriages in hopes of preventing difficulties. Others come for help after many years of marriage. In any event, there can be no doubt that dissatisfaction with the current state of a marriage does not spring full-blown out of a single evening's boredom or a single bad experience, no matter how traumatic. Nor does the desire to find greater satisfactions in marriage and the attempt to seek ways of increasing them appear overnight.

For some couples, casework marriage counseling is the answer. In other cases, one of the marital partners may need medical treatment by a psychiatrist while the spouse is seen by a caseworker. Current counseling practice in our agency, discussed in the second section of this chapter, incorporates several techniques.

Increasing emphasis on the treatment of marital problems as a central focus brought with it more intensified examination of diagnostic factors. Knowledge derived from psychiatry became an integral part of the operation. It was found that an assessment of the interpersonal relationships, personality traits and characteristics of the marital partners, as they combine either to support each other or produce conflict, is essential to the practice of marital counseling.

The training program for caseworkers has as its objective the deepening of the worker's ability to see and understand the family members as they are; as they regard, use and respond to each other; how much in that response to and use of the other is based on reality, and how much on neurotic distortion, social dictates or cultural attitudes.

Continuous training seminars that are geared to the level of experience of the workers are offered by case supervisors and psychiatrists. Individual supervision has been and continues to be the basic training method in casework. Supervision is concerned with the integration of theory and practice in the individual worker. While intellectual comprehension and emotional sensitivity are vital in casework, their translation into skill and method is a crucial aspect of training.

The worker needs to know what kind of information to obtain, and how it must be obtained in the treatment relationship. He must learn how to evaluate data, how to test for accuracy in the midst of the highly charged feelings that usually accompany the description of conflictual material. He must learn the pace at which it is safe to move, and the effect his own personality has on the client and the relationship. Further, he must learn to hear the unverbalized implications that are manifest in the client's attitude, his selection of problem, and the like. These are but a few of the many essentials in the interaction between worker and client.

As mentioned earlier, in the treatment of marital problems, in addition to the individual personalities involved, we are concerned with the interaction between marital partners as a significant factor in appraising the degree of family stability. Where a caseworker is carrying both partners in a marriage, this requires an ability on the part of the worker to relate to each of the partners separately without becoming over-identified with either. While this is more taxing than working with only one member of a family, it often leads to good results.

The use of a psychiatrist's services for a diagnostic evaluation has been integral to family agency service over a great many years. Only recently, however, have we been able to define clearly the most productive collaboration of psychiatry and family-oriented casework treatment. Both disciplines—casework with its experience in the family's social disturbance, and psychiatry with its experience in the individual's intra-psychic disturbance—are brought together to combine their knowledge for the benefit of the family unit. Essential to treatment planning is a complete understanding of individual personalities and their interaction, and the extent to which the social setting supports, aggravates or lessens the tensions and strains. In considering possibilities of changes in the interactive patterns, decision needs to be made with regard to the extent to which changes in the social situation can be effective instruments for altering the demands each is currently making on the other, and the extent to which the marital partners will need to adjust to the demands which the social situation will continue to make. The close collaboration between psychiatry and casework is made essential insofar as no single discipline in these specific instances can supply a complete understanding of all the factors in the marital conflict, and how these operate in and affect the family situation.

For psychiatry to make its fullest contribution in this relationship, it is essential that the cooperating psychiatrist understand agency function, practice and goals. For casework to make its fullest contribution to this relationship, it is necessary that it understand the psychiatric contribution from the vantage point of that discipline and assume its own responsibility for the integration of this contribution into casework practice, treatment plan and goal. While this collaboration has broadened the horizon of each discipline, its value has rested in the assumption by each of final responsibility for practice in its own field.

In the past, when one of the members of a family required psychiatric treatment we were faced with the problem of inadequate community treatment facilities and the difficulty of arranging a close collaborative relationship. Even with the best intentions, this usually resulted in two separate, mutually exclusive processes with the same client.

A number of distinct stages preceded the emergence of our current methods. In the early thirties there was generally one caseworker for both partners in a marriage. In certain situations two caseworkers would see the two partners. The forties brought a greatly increased use of the psychiatrist in consultation. A considerable number of people were referred by the agency for private psychiatric treatment where it was indicated. Since World War II the psychiatric program has been expanded not only in consultation but in teaching in our agency as well as in others throughout the country.

By 1953, the increasing collaboration of the psychiatrist on the staff of our agency with our supervisors and workers extended beyond consultations to a demonstration of concurrent casework-psychiatric treatment in cases where the marital problem is complicated by a degree of disturbance in one of the partners that contraindicates casework help for him. This program of concurrent treatment of a couple by psychiatrist and social worker is now in its third year of operation and already shows much promise. Twenty to twenty-five couples a year have been seen during this period, one of a pair by the psychiatrists, the other by a caseworker, with frequent consultations between the two. Through this close association, both learn a great deal about the minutiae of interaction in a particular pair.

Where indicated both partners are screened by a psychiatrist. Determination is made as to which partner should enter psychiatric treatment and which should continue with casework treatment by a joint psychiatrist-caseworker consideration of the dynamics, the social situation and possible goals in treatment. The caseworker is guided by the psychiatrist's interpretation of the dynamics in the case. While the project utilizing this technique is numerically small, it is of enormous proportions in its implications for the future. Directions for research in this collaboration are outlined in the following section.

II *Present Status of Treatment Program*

M. ROBERT GOMBERG, Ph.D.*

CASEWORK TREATMENT of marital problems represents a substantial part of the caseload of family agencies across the country. In the Jewish Family Service of New York, for example, of the 3,639 families that received extended casework services in 1954, 2,799 or 79 per cent had marital problems. Casework, in close conjunction with psychiatry, has been evolving a modality of treatment which holds much promise for the innumerable marriages that require such help.

A thorough presentation of the treatment process in work with marital problems in a family agency would require an examination of such basic casework assumptions and concepts as the need to understand character structure, personality development and the dynamics of interpersonal relations, as well as the relation of these psychological factors to psycho-social diagnosis, to the casework method, and to differentiated treatment goals. Obviously, it is not possible to present so comprehensive a review in this chapter. Our purpose here is to discuss the special aspects of family-oriented treatment of marital problems in a family agency.

Family diagnosis and family-oriented treatment, by definition, throw the spotlight on the interaction between significant members of a family; in a marital problem, on the interaction between husband and wife. Although a sound clinical understanding of each partner is essential, it is equally important that the relationship between the partners be understood. This interaction should be viewed as a vital phenomenon, crucial to the marital adjustment, and therefore crucial in treatment.

---

* Executive Director, Jewish Family Service, New York City.

Only recently have we focused attention on the interaction in marriage, recognizing it as a separate factor—something beyond the intrapsychic phenomena of the individual personalties involved. It is interesting to note the development of the literature on the subject of reciprocal influence.[1-5] In essence this literature stresses the fact that new knowledge must be added to the existing body of information about personality development, motivation, change, treatment techniques, and so on. It also points up a need to reconsider and possibly correct certain assumptions on which we have based our practice. If this impression is correct, what assumptions require re-evaluation and revision? The basic one, we believe, is that the quality of a marital relationship is only a by-product of the degree of health or disturbance in the personalities of the two partners.

From the traditional position, the marital relationship is viewed on a one-dimensional screen showing only the healthy and/or neurotic behavior of each partner, intact and unchanging, with minimal appreciation of the "new compound" formed by the interaction between personalities. This view does not show a second dimension—the use that each partner makes of the other.

This latter orientation has considerable significance from both the social and clinical points of view. Socially, we are all aware of the great incidence of family disruption, marital discord, separation and divorce. We are concerned with the impact on children, whether evidenced in quiet, unreported suffering or in the glaring headlines on deliquency. Clinically, there are two implications. First, a "good" marriage and a "good" family do not have as the ideal prerequisite two neurosis-free individuals, however desirable that may be. The constructive and destructive elements and their combination in the interaction are at least as critical to the ultimate balance or equilibrium attained in the marriage. Secondly, it is possible to offer treatment for certain discordant marital situations, to help achieve substantial improvements in these relations, without working through all the unconscious neurotic complications in each partner.

The selection of the family for marital counseling or casework treatment must be made on a sound diagnostic basis. Many marital situations reveal the kind and degree of pathology and destructive interaction that contraindicate casework help. In these, referral for psychiatric treatment is necessary. In some situations, although the problem is described as marital, it is a displacement. The person applying soon reveals the degree of ego impairment and character disorder requiring psychiatric attention. In such instances, however, we make every effort to offer or get help for the other spouse, even though he did not apply originally. If completely left out of the therapeutic experience he may continue to act out in the marital relation behavior which is destructive or retaliatory. He may grow hostile to his partner's treatment and consciously or unconsciously obstruct or sabotage it. With casework treatment, it is possible for him to gain some understanding of the partner's illness as well as his own role in the problem. Thus, supportive responses may be motivated in him.

However, there are countless families for whom casework, with psychiatric diagnosis and consultation as needed, is the appropriate treatment. The goal of casework is not character change or the treatment of pathological processes. Casework deals with the adaptive functions of the ego. When there is sufficient ego strength or maturity, with sufficient constructive interaction despite the disruptive forces at work, casework is able to build on the strengths within the family gestalt.

We have seen great numbers of disturbed marital relations where tensions, conflict and distortions were mounting and threatening the stability of the relationship, but which were accessible to working through channels of ego clarification of attitudes, feelings and reactions. This led to more effective self-understanding, understanding of the mate and the relationship, and understanding of the role in marriage. The result was more effective functioning and a more productive equilibrium in the marriage. Such results are predicated on sound diagnosis, case selection, and defined and focused goals of treatment.

Psycho-social diagnosis of a marital problem requires a careful clinical picture of the separate personalities; a psychological evaluation of the relationship; an assessment of the degree of readiness and motivation to use help, and is also influenced by social, cultural and economic factors.

The case history which follows was chosen to illustrate casework treatment of marital problems for several reasons. First of all, it illustrates concurrent treatment of both husband and wife. Secondly, it demonstrates clearly the complementary nature of the personality problems, as well as the negative and constructive elements in the interaction within the marital relationship. Thirdly, in both these partners one notes the conflict between unresolved personal problems stemming from earlier family life and counteracting healthy ego drives and personality strengths.

Mrs. Brown, an attractive young woman of twenty-eight, applied to the agency for help with her marital problem which had reached a point of desperation. As a schoolteacher, she knew something about the services of our agency. She had been married for three years and had known her husband for three years prior to the marriage. She thought of him as strong, reliable and intelligent. During their courtship the relationship had "matured and developed," and she had looked forward to their marriage.

Their difficulties began when they had been married for about a year and a half. Mrs. Brown told of an increasing number of bitter quarrels and arguments with long periods of angry silence in between. Narrowing the area of her complaints, she focused on what she considered her husband's overly cautious, unspontaneous, methodical behavior. While this had not offended her earlier, she now found herself growing increasingly tense and exasperated by it, since she herself was "spontaneous, outgoing and vivacious."

While the contrast between them had originally seemed a healthy "combination of opposites," these differences were now keeping them in perpetual conflict. The satisfactions in their marriage had drained off; their sexual relationship had been affected; and the many values and gratifications she had known earlier in the relationship had slipped through her fingers. She could not see herself going on unless some change took place.

Interestingly, her behavior as first observed in the interview seemed to be just as she described it in her marriage—diametrically opposed to that of her husband. She seemed a very impulsive, highly emotional young woman, yet there was apparent over-determination in her enthusiasms. She worked so hard at whatever she did— laughed too hard, cried too hard. She had an obvious need to play each emotion to the hilt. Thus, in her own way, she lacked the very spontaneity she regarded as being absent in her husband.

Briefly, Mrs. Brown's background was as follows: She came from a lower middle income family where the father was in business and had provided modest economic security. Both parents had some education and were interested in cultural activities, particularly in the arts. The oldest of two children, with a brother four years her junior, Mrs. Brown described her family as a warm, close-knit unit with many interests in common, a lot of family pride and family feeling. They were always proud of her good scholastic record and her many social successes. Other relatives envied their closeness.

Actually, however, when we began to examine more closely the mother-daughter relationship, the father-daughter relationship, the sibling relationships, she reported a great many anxiety-producing experiences which inevitably had some impairing effect on her personality development. An appraisal of only these unilateral relationships and their potentially harmful effects would, in fact, lead to the conclusion that this girl's ego had been most seriously impaired. We therefore had to account for the degree of personality development she had achieved. It became clear, as we looked more closely, that within each of these separate relationships there had been certain supportive and constructive elements. Even more important, it was also evident that there had been nurturing qualities in the total family interaction that had contributed to her growth.

Mrs. Brown described her mother as an attractive, intelligent and highly self-centered woman. Her outstanding qualities were, first of all, her self-centeredness and, secondly, a constant ailing and hypochondriasis that went back to Mrs. Brown's earliest years. Mrs. Brown recalled being told by her mother that she had almost died when Mrs. Brown was born.

Her mother seemed to bring subtle rivalry and envy as well as enthusiastic interest and pleasure to each meaningful incident in Mrs. Brown's development. For example, when they moved to a "good" neighborhood and the school was a "good" school, her mother expressed her great happiness that the change had been possible, but then proceeded to tell countless stories of her own impoverished background, her own lack of adequate housing and school facilities. The same ambivalent reaction on the part of her mother took place whether the event had to do with neighborhood, school, friends, clothes, dates, clubs, or other group activities.

Two points are of interest. First, Mrs. Brown had been successful in her social life and in school. She always had had several close friends and a great many acquaintances, and had taken the leadership role. She had been a good student through grade school, high school and college. Mrs. Brown had enjoyed these experiences and gained a great deal from them. At the same time she was aware of a sense of guilt toward her mother. Quite early she had developed a feeling that somehow these pleasurable experiences by right belonged to this woman, who had been deprived of gratifications in her own youth. Thus, even as Mrs. Brown participated in them, enjoyed them in part, found success in them, gained the approval of her peers and elders, a feeling lurked in the background that by accepting these experiences for herself she was continuing her mother's deprivations.

It became clear that the so-called "spontaneity and impulsiveness" were essentially parts of a reaction formation; by plunging into experiences she could not be charged with deliberately having chosen activities that "properly" belonged to mother. Her mother's partially encouraging attitude about these experiences and Mrs. Brown's own satisfaction in them served to further her growth and development. But since the mother's attitude also engendered hostility and guilt, which Mrs. Brown had had to repress, her maturational process had been blocked and her self-image and values had become distorted.

Mrs. Brown described her father as a mild, passive person, completely controlled by his wife. It was only after careful review and discussion that Mrs. Brown was able to reveal the extent of her own positive feeling for him. There was a combination of deep affection and great disappointment. Very early in life she found she could not rely on him. Although she sensed that he sided with her when she got into any kind of difficulty with her mother, she could never count on his support. There was a feeling of something clandestine in her relationship with him. He never emerged as a strong parental figure within the home, and her affection for him, though strong, was somewhat obscured by this fact. Thus she once commented, "Men are nice, but you really can't quite rely on them."

Again, in discussing her brother, there was a marked duality of feeling and experience. On the one hand she described a warm, close sibling relationship through the years, with full recognition and acceptance of the inevitable rivalry and fighting, with amusing stories of her annoyance with him during her teens when he tagged along, and many other homely incidents—all told with affectionate overtones. On the other hand it became quite clear in the course of the discussions that from the earliest days she felt poignantly that he had displaced her in her parents' favor, particularly with the mother who made her preference for the boy rather obvious. Like herself, the brother was an excellent student, winning many honors in both high school and college, and is currently engaged in a successful professional career. They continue to meet frequently with much family visiting to and fro.

The significant attitudes that emerged were Mrs. Brown's resentment and fear of having been displaced by her brother in her mother's affections, her guilt in relation to her mother and her insecurity because of the unreliability and unsupportive quality of her father's affection. As a result of these factors, quite early in life she had started to woo people, trying to attract attention, admiration and love at any price. She needed to recapture the feeling of "coming first" with all its implications. Her efforts to attract people now were motivated not only by a desire to fulfill natural affectional drives, but were also motivated and accompanied by anxiety and guilt. Because of her repressed resentment, hostility and guilt, her efforts to attract interest and admiration even when successful were only partially satisfying and failed to provide proportionate growth-inducing experiences.

At home, in her marriage, with friends, in the community, on her job, with her colleagues, the over-all picture was that of a woman respected, admired, looked up to and sought after. Undoubtedly, this afforded her some satisfaction and gratification which tended to support and strengthen the healthy ego tendencies within her. On the other hand, there was an evident imbalance between the energy she expended and the amount of inner security she achieved.

Mrs. Brown was seen for a period of about two months before her husband began his treatment experience. Treatment continued for both of them for a period of one year.

The husband presented the following picture:

Mr. Brown is thirty years old, a chemist, and the only son of well-to-do parents. His father is a moderately successful attorney. Here, too, the family was described as being rather "close." There was no open friction and there had been a great deal of joint family activity. Always a "family man," the father had taken his son to ball games, movies and the like. However, he was a most exacting taskmaster, a perfectionist when it came to such matters as cleanliness, morals, marks and achievement in school.

Mr. Brown's mother was a very ambitious, driving woman with elaborate plans for her son, and when he was quite young he sensed that in order to find a secure place and earn the acceptance of his family he must produce honors. He would be loved or judged less for himself as an individual and more for his performances and achievements. Thus, success in school, in sports, in relationships with other youngsters reflected a pressure upon him beyond the personal gratifications achieved. The experiences of being a fine student, a good athlete, an active member of clubs, athletic teams, scouts, and the like were meaningful for him. But a good measure of the value of these experiences was diverted as a security offering on the altar of his parents' ideals. In this way he felt he could hold on to what seemed to him a precarious tie with their affection.

This situation was exemplified in an experience he recalled having at the age of seven. While he was doing his arithmetic homework, his father's arm shot out from over the boy's shoulder, picked up the paper and tore it to bits. The resentment and hostility of this act were not due to the fact that the boy's work was incorrect, but because it was sloppily written and some numbers had been placed below the line. The incident indelibly impressed upon the boy the standard demanded by his father and the penalties to be expected in the event of failure to live up to his expectations.

Mr. Brown's mother was constantly comparing him with his contemporaries. Was he taller, stronger, smarter? Were his grades better? Did his teachers prefer him to the others? Whatever he did was under the closest surveillance, with the result that he always had a feeling something more was expected of him. That both parents loved and admired him was undoubtedly true. In spite of all the expecting pressures placed upon him there were some rewards that served to meet many of his growth needs. Yet because of the pressures, demands and expectations, he could not inwardly meet the excessive demands of his own incorporated standards. His pattern was to move slowly and cautiously into new experiences, and to calculate all risks. He wanted to know what he was getting into and what the chances were of succeeding. However, he was never immobilized, and was able to take on responsibility for decisions and to follow through on them.

It was Mr. Brown's steady, clear-thinking, well-organized approach to life that had first attracted his wife. This compensated for a part of herself that was lacking. Mr. Brown, in turn, was drawn to the outgoing, spontaneous qualities he thought he saw in Mrs. Brown. Because of the degree to which the qualities had been integrated in their respective egos and the support and consolidation they received from reality successes, these respective patterns were not merely defenses against their repressed individual anxieties, they actually were evidences of strong capacities.

During their courtship and at the beginning of their marriage, Mr. Brown relied heavily on his wife's social grace. It was she who broke the ice and made new friends. With her help he found it much easier to communicate with people and participate in social events. For her part, Mrs. Brown was comforted by the thought that his steadfastness would serve as a check against her "going too far." She was pleased that their friends and acquaintances regarded him as a reliable person to whom they could turn for advice. Each counted on these respective qualities in the other, leaned on them and was supported by them.

While one must be alert to the defensive component of these patterns, the constructive values and ego strengths represented in them cannot be dismissed. In casework treatment we attempt to draw upon, build upon and enlarge the scope and flexibility of the healthy aspects of ego tendencies. An understanding of the negative component helps to determine what should be handled directly and worked on in treatment and what should be avoided, in keeping with casework methods and objectives.

In the early phase of Mrs. Brown's treatment, there was a careful review of the complaints and conflicts she described. She gave innumerable illustrations of her husband's behavior which irritated, embarrassed and made her anxious and hostile. As she was helped to elaborate, clarify and identify her feelings regarding these provoking incidents, a thematic pattern emerged. The anxiety and hostility were usually not initiated as a result of Mr. Brown's behavior toward her personally. The events that precipitated crises always involved one or more other persons, as, for instance, when they went to the theatre with friends. During intermission there would be an exchange of interested and critical appraisal. Mrs. Brown said that when her husband was asked his opinion he would comment that it was only the first act and he wanted to wait and see the rest of the play before expressing his views.

Again, when discussing politics with friends (most of whom seemed to think alike), her husband would always imply that things weren't strictly black and white and that perhaps some consideration ought to be given to "the other points of view." Thus, while Mrs. Brown would pick up the feeling tone of the group, move with it and become most articulate in expressing their view, Mr. Brown wanted to mull things over and examine all sides before committing himself.

The Browns would take each incident home with them and relive it over and over again. She would chastise him for alienating people and losing friends (an accusation which was completely unfounded). Then she would grow bitter and hostile and lose control. He would try to argue and then withdraw into a sullen silence, going into the other room to "work." His silences infuriated her and at the same time made her increasingly anxious.

As Mrs. Brown grew more secure in the treatment relationship, it became clear that essentially she did not disagree with her husband. More often than not, she shared his opinion about the theatre, politics, community activity and the like. Her own pattern of being most anxious about the group attitude, of needing to conform to the group in its judgment of any matter was clearly distorted out of all proportion, since it was evident that the Browns held an important place within the group and were highly regarded.

Gradually Mrs. Brown began to understand the extent to which she had been projecting blame onto her husband, while in reality the anxiety she was experiencing came from other sources. She continued to be irritated and annoyed with her husband, but there was some lessening of the degree and intensity of their difficulties. Her interest began to shift from a constant reiteration of attacks upon him to the question of why she was so upset. Why did she feel so insecure with the group that she was left with little room for her own feeling and judgment?

During the first two months, before her husband entered treatment, Mrs. Brown was warm, grateful and responsive to the worker. She seemed to be making progress in her relationship with her husband, although she "still blew her top" from time to time. Shortly after he began, however, there was a setback in the marital relationship and a change of attitude in treatment. An edge of anger crept into her tone during her interviews. She began to build up indictments against her husband again. She was overly sensitive and easily hurt by observations from the worker which she had previously accepted with interest and thoughtfulness.

When the worker pointed out the undertone of anger and sensitivity, Mrs. Brown at first protested and said she was really grateful. But when the worker suggested that this might give some clue to the question of why she displaced so much of the feeling onto her husband without really knowing or facing what it was that was troubling her, she acknowledged that she had been aware for several weeks of some resistance to coming for treatment. She admitted a reduction in her high hopes and enthusiasm about what help could do for her, and even a feeling of anger against the worker.

In examining this change of attitude, it became clear that the turning point coincided with her husband's entering treatment. The worker had been a kind, protective, guiding parent symbol, interested exclusively in her. Her husband's coming into treatment was once again the threat Mrs. Brown experienced in early life when her brother arrived on the scene and displaced her in her parents' affections. Now the worker had become the father who never backed her up, and once again the fear of "being pushed out" made Mrs. Brown behave as though it had really happened.

By helping her see these connections—by comparing the present experience with the background she had reported earlier and which she now elaborated in greater detail—the worker treated this as a reality testing experience. He reassured her that his attitude and concern in helping her had not shifted at all, that he was still very much concerned with the over-all purpose of strengthening and working out her marital relationship.

In examining the feeling she had expressed in relation to the worker, Mrs. Brown realized the extent to which, in the beginning, she had been extremely warm and positive, perhaps overly so. This was partly due to her relief in getting help and being able to express some of her difficulties. She had also put to work the same pattern of "needing to be liked" at any cost. Later she discovered by actual experience that her security did not depend upon this, that her attitudes and feelings were accepted, and that she could agree or differ with what was said without jeopardizing this relationship. This realization had important therapeutic value.

As for the fear of being displaced which had apparently precipitated the serious complication in her marriage, the perplexing thing was that she had not been troubled by it in all the previous years. It was only during the preceding year and a half that the difficulties had risen to the surface. She pointed out that the trouble began just about the time the couple was seriously considering having a baby. Before that they had discussed the matter in rather general terms of "having a family when we're ready." However, pressure from their parents, the fact that many of their contemporaries were having children, and the fact that their own economic position presented no problem were making them face the issue.

From the beginning Mrs. Brown took the position that it was she who wanted the child, and increasingly accused her husband of "a lack of enthusiasm." At the same time he insisted he really did want a child and just didn't know how he could "prove it to her." In treatment Mrs. Brown came to realize that she said she wanted a child partly because "it was expected of her" and partly because she really did. Yet she was very much afraid of the idea and had projected this fear onto her husband.

Here again her feelings were examined and clarified on a reality-oriented basis. As a result she was able to face her own ambivalence with genuine relief. She realized that the image she had built up of herself and the importance she placed upon her social role and position left no room for doubts—certainly none about being ready for or wanting a child. Perhaps most significant was her recall of and association with the last time a baby "had come into my life"—her brother. At that time the event had had what seemed like a depriving effect upon her.

Mrs. Brown reported with satisfaction that she was discussing with her husband much of what we had been talking about and that he was sharing with her some of his own new understanding of himself. It seemed to bring them closer together again. The wall that had been build up between them appeared to be giving way. Their social as well as sexual relationship was greatly improved. Mrs. Brown realized that her reservations and uncertainties about having a child, along with her inability to face and express the problem had led her into hostile and repudiating behavior toward her husband. As the potential father of the child, he would be instrumental in creating a situation within her new family where once again she might be displaced by the child.

This fear of being rejected spilled over into many of her significant relationships—first with her husband, then with friends and with the worker. In attempting to protect herself against this anxiety and secure her place with people who had value and meaning for her, the unhealthy component of the "spontaneity" of her pattern came into exaggerated play. She wanted to make sure she was not losing her place in the feelings of others, hence her efforts to sense what her friends thought or felt and then to become an indispensable part of those feelings and thoughts.

During this time her husband's conservative pattern became a threat to her. While she had counted and leaned on him in the past, when he ceased to operate in the image of her own defenses, his behavior represented some danger. Realizing this fact, Mrs. Brown commented that she had really been "cutting my nose off to spite my face." By not permitting him to be an individual with his own reactions, his own thoughts and feelings, she was actually depriving herself of what had been an important source of strength. (In marital problems we frequently find the Pygmalion motif, where one partner seems to wish to remake the other person's personality. In reality he needs the marital partner to serve as a bulwark to strenghen his own defenses and sees the mate less as a separate person than as an extension of himself.)

Throughout Mrs. Brown's treatment we worked with those feelings and attitudes that were readily accessible to her—her fear of rejection and displacement, her lack of readiness to have a child, and her inability to face and acknowledge this fact. In dealing with the complex of these feelings and their effect on the marriage, she was helped to dispense with the presenting screen of distortion and misconception. She was guided in a recognition of where she had some justification for her resentment and the extent to which she had exaggerated her feeling of being unloved. This was meaningful in her relationship with her parents, which she was now able to appraise on a more realistic basis—expecting and demanding less and therefore enjoying it more.

A number of points in Mr. Brown's treatment are interesting. When he first came for treatment he was again being the "good and cooperative son." Completely confused and bewildered by what had happened to the marriage, he was prepared to do anything to improve it. His wife's behavior had reawakened his

earlier experience of having someone constantly expecting something of him, and his own feeling of never being quite able to
fulfill the demand. He was completely unaware of the fact that
he resorted to punitive and retaliatory behavior. It was evident
that he had a genuine love for his wife, but what he had been
unable to acknowledge to himself was the extent to which her
difficult and provocative behavior had engendered resentment
and hostility in him.

Little by little he recognized that his tendency to "withdraw,"
to find he always had some important work to do that prevented his continuing a discussion, was his indirect way of
punishing her. He also became aware of the extent to which his
feelings were colored and exaggerated by his earlier experiences
in his parental home.

Mr. Brown was helped to clarify his own feelings, to affirm
those that were truly affectionate and concerned for his wife,
to face his resentment and hostility, and to distinguish between
his real difficulties with his wife and those resulting from displaced resentment against his parents. As a result he became
more effective in the marital relationship and grew more toler-
ant of his wife's moods.

Both had reported that their sexual relationship, previously
mutually satisfying, had begun to deteriorate during the discordant period of their marriage. Here again Mr. Brown was
able to examine his role in the situation and then correct it.
Because of his wife's disturbed and sometimes depressed feelings,
and his unconscious tendency to let her take the lead, he had
permitted the frequency of intercourse to drop off, and the satisfaction in the experience itself was reduced. On reviewing
this with him, he was encouraged to assume more initiative. His
greater ability to tolerate and understand his wife's feelings
plus his new way of asserting himself in their sexual relationship
led to improvement in this aspect.

Another factor of great importance to Mr. Brown was the
over-all question of leadership. Although he complained mildly

about his wife's need to be so active socially, it was apparent that he left her all decisions as to the selection of friends, interests and activities. As he became more aware of his dependence on her in this respect, he started to assume more responsibility for participating in the choice of social activities and relationships.

Obviously, Mr. Brown was less troubled than his wife. Had the crisis in their relationship not developed when it did he would have been content to keep on functioning as he had. The little satisfactions he derived from marriage, from his work and other activities seemed sufficient for him. However, the treatment that followed the crisis served not only to restore the marriage to a more effective state, but increased gratifications, his own self-understanding and, in some measure, contributed to his further maturation.

As noted earlier, Mr. and Mrs. Brown were seen over a period of a year—the wife for fifty interviews, the husband for forty-one. By the end of treatment both appeared to be in effective control and functioning on a level that seemed to meet their respective needs and to restore a healthier balance between them. Each still had a certain unresolved, unconscious neurotic core, but by drawing upon the respective strengths in each personality, as well as on the interaction between them, they now enjoyed a considerably sounder relationship.

Mr. and Mrs. Brown were also seen twice for follow-up interviews at intervals of one year in the two years after termination of treatment. During this period a baby girl was born to the couple, and when last seen both Mr. and Mrs. Brown reported a sense of well-being and stability in their family relationship.

We believe that because of emotional interaction in marriage it is logical that treatment, wherever possible, be made available to both partners. This raises some technical questions that require further exploration: For example, the matter of proper timing for involving the second partner of the marriage in treatment. While experience demonstrates that, by and large, it is sound to see both partners, there are times when it is contraindicated.

Further, there is the question of whether one or two workers should carry the case. At present we lean toward the practice of one worker carrying both partners, unless there are sound indications to the contrary. While it is technically more difficult and complicated, there are many advantages. In the Brown case, for instance, the wife's reaction to her husband's beginning treatment with the same worker was of utmost significance. The worker's ability to help her see that the shift in her own attitude and feeling did not stem from any change in the reality of the treatment situation had considerable therapeutic value. However, there are many situations where it is wiser to have two workers carry the case. In such instances, of course, there must be the closest collaboration between the workers.

A psychiatric consultation in relation to one or both marital partners is frequently utilized. The psychiatric diagnosis may help to determine whether the case is suitable for casework treatment or requires psychiatric treatment. The formulation of the dynamics of the case, when psychiatric consultation is required, is helpful to the caseworker in the approach to either or both partners.

It cannot be stressed too often that casework treatment of marital problems must be based on a psycho-social understanding of the individual members of the marital relationship and their personalities and of the joint interactive relationship which functions as a "third force" in the evaluative process. Each of the marital partners possesses personality traits and defensive patterns which operate within a wide range—from complementary to antithetical—in the marriage. The degree of marital equilibrium is not necessarily based on the degree of emotional health of the individual partners.

In casework with marital problems, the arena for treatment of each individual is the adaptive functioning of his ego. Casework treatment enlists those capacities for motility and object relationship of the individual ego that are not pathologically impaired. Not least among the factors drawn upon is the ego ideal of positive family life and marital relations of the individual and his drive to achieve them.

Research Implications

At their inception, casework developed services, as we have seen, learned to manipulate the environment in support of individuals and families, and later, with new insights into dynamic psychology, integrated into its conceptual system and practice a balanced appreciation of the outer and inner forces that must be brought into some kind of harmonious relation if man is to function as a social being. Diagnosis and skills were evolved to deal with both of these components wherever and whenever casework could provide the appropriate treatment.

The present method and philosophy of casework treatment was arrived at as a result of experience with many hundreds of thousands of individual and family situations. Very early in its development, as previously mentioned, casework developed a system of control through supervision, building into its practice a continuing form of self-evaluation and criticism, as well as training. By so doing we were able to distill this vast experience and arrive at hypotheses and techniques of practice based on empirical judgment and evaluation of results. In addition to this continuous process of evaluation as an integral part of any agency operation, there were the determinative contributions of the important leaders in the field that helped to crystallize thinking and break new ground for progress.

Today, the field of casework has begun to look to systematic scientific research as a new instrument for both critical examination and testing of its assumptions and hypotheses and for the further development of methods of diagnosis and treatment. For a long time it was believed that the treatment relationship focused on personal and social adjustment was too dynamic and elusive to afford a field for research investigation. Judgments based upon empirical experience, supervision, self-evaluation and criticism were regarded as the only tools for study. This in turn meant that an attempt to validate hypotheses scientifically was virtually impossible.

However, in the past decade or so we have come to recognize that new research techniques are in process of being fashioned, so that the kind of data with which we deal will become more accessible to scientific study. A few family agencies, the Jewish Family Service included, have instituted departments of research. In our own agency we use a team composed of a research psychologist, a number of research fellows, a casework research associate and psychiatric consultation. Even in a relatively short time we have found a variety of values with immediate usefulness as well as potential contributions toward the long-range goal of more exact knowledge and understanding of personality, behavior, adjustment and treatment method.

Our research department has completed two studies: the first deals with the effectiveness of "brief contact"; the second with a study of follow-up interviews. The new project which the department is currently investigating is in direct line with the subject of interaction in marriage. Empirically, we have discovered that the degree of presence or absence of neurotic behavior in one or both partners in a marriage is not the only determining factor in appraising the stability or instability of the family as a whole. The need for additional diagnostic classifications becomes apparent. It has been our experience that there are families in which the husband and wife are relatively free of personality disorder and yet combine to make a poor marriage. Conversely, we have seen literally thousands of families where individual diagnosis reveals neurotic elements and yet the combination results in some stability in both the marital and parental roles.

If the existing diagnostic tools and classifications are equipped to measure only the nature and type of individual personality with which we are dealing, the tendency is for treatment to follow such diagnosis and attack the neurotic component of the individual. Obviously, this is important and must be part of any diagnostic work-up. However, the present study will attempt to delineate the kinds of combinations of personality which, in spite of neurotic components, combine to achieve effective marital adjustment, as against those which combine to result in breakdown of marital relationships. This data will be compared with clinical diagnoses to see what significant groupings are revealed.

The foregoing is obviously a very general and broad description of what is worked out in great detail in the research design, with careful consideration for groupings of "stable" and "unstable" families, criteria for these judgments, scaling techniques for evaluation, etc. Perhaps of the greatest importance is the effort to attack this kind of problem through research, which we believe and hope is developing the skills for such measurement. The more sensitive, reliable and valid the research technique becomes, the more penetrating is its capacity to examine critically and evaluate hypotheses which have grown up at times less from adequate evidence than from repetition and tradition. Even if this creates certain problems in the process, the rewards in increasing the scientific premises of our practice are obvious.

Family agency experience, now of considerable proportions, has demonstrated that when the psycho-social approach is used —i.e. when there is diagnostic understanding of the individual, a proper evaluation and understanding of the quality of interaction, and a primary determination of the suitability for the psycho-social approach—casework offers an invaluable modality of treatment for marital problems.

References

1. Ackerman, Nathan W., "The Diagnosis of Neurotic Marital Interaction." *Social Casework*, XV, No. 4, April, 1954.

2. Bychowski, Gustav and Despert, Louise, eds., *Specialized Techniques in Psychotherapy*. New York, Basic Books, Inc., 1952.

3. Gomberg, M. Robert, *Family Counseling, Practice and Teaching*. New York, Jewish Family Service, 1949.

4. Green, Sidney L., "Psychoanalytic Contributions to Casework Treatment of Marital Problems." *Social Casework*, XV, No. 10, December, 1954.

5. Pollak, Otto, "Relationships Between Social Science and Child Guidance Practice." *American Sociological Review*, XVI, No. 1, February, 1951.

CHAPTER XVI

Group Approaches to the Treatment of Marital Problems

———•◦•———

I Group Education

GERTRUDE GOLLER*

IN RECENT YEARS there has been a growing emphasis on the use of educational methods to prevent marital problems and to strengthen family relationships. High schools and colleges throughout the country have an increasingly large enrollment in courses dealing with "Family Life Education" or "Preparation for Marriage." Newspaper columns, mental health films, lectures, radio and television programs are among the mass media used extensively to educate for family living. The general or specific goal in each of these programs is the prevention of marital breakdowns through increased knowledge and understanding of what is involved in marriage, and support of the constructive elements in a marriage.

The focus here is on small group programs primarily for en-

---

* Associate Director, Department of Parent Group Education, Child Study Association of America.

gaged couples or parents, with limited and constant membership and provision for at least several sessions; the method is primarily one of group discussion. Such programs are being conducted under the auspices of adult education centers, churches, group work and family casework agencies and colleges, and lead by social workers, psychologists, teachers, guidance counselors, physicians, adult educators and religious leaders. The programs do not purport to cure problems caused by emotional conflict. Rather, they aim to prevent or alleviate discord caused by lack of information, inexperience, or attitudes which are modifiable by the insights and experiences gained through the content and method of the educational group.

The groups for engaged couples are organized by several agencies and particularly by planned parenthood centers. Usually they are limited in size to about six couples who meet with a professional, trained leader three times a week. Often the group members determine the topics for discussion. Among these are early emotional adjustment in marriage, sexual adjustment in marriage and family planning. The many groups meeting under church auspices are likely to include discussion of the spiritual aspects of marriage.

Whatever the auspices, the method used in conducting most of these groups is one of combined lecture and discussion. The leader or speaker gives a talk on a specific topic. The members then have an opportunity to raise questions, share common concerns and experiences, and to explore their own attitudes in the particular area under discussion and their affect upon their marital relationships. Agencies conducting these programs report that many who enroll in these groups are referred by others whose needs have been filled by such sessions.

In the educational groups for parents, the focus is on parent-child relations, rather than on the marital relationship per se. These programs, like those for engaged couples, are considered educational in nature and preventive in purpose.

Formal parent education programs have been in existence for

well over fifty years. Although their method and content have changed over the years, their primary purpose has remained constant—to help parents gain increased understanding of child development in general and of themselves and their children in particular.

The typical group described here has a professional, trained leader and uses group discussion as its primary method. Such groups, existing in many parts of the country under a variety of auspices, are generally composed of parents whose children are in a particular age range, sometimes of a few years (teen-age) and sometimes over a wider span (six to twelve years). The group contains from ten to twenty members who are expected to enroll for the entire series of eight or more weekly sessions.

The purpose of such groups is to afford parents an opportunity to discuss their common everyday experiences with their children (as opposed to offering help regarding long-standing or serious disturbances in a child's behavior or the parent-child relationship). As one parent discusses his own child, the leader encourages the others to share their experiences in similar situations. There is discussion of feeling as well as of fact. From the variety of experiences and attitudes of the group members, the leader helps them gain an understanding of the fact that each child has his own pattern of growth, and that parents respond differently to children's various developmental phases.

Despite the fact that the purpose of this kind of group is primarily educational, members may enroll who want and need help for uncommon situations. When they have had an opportunity, through group discussion, to see the more usual range and variety in children's behavior and in parental reactions, parents with particular problems may then come to realize that they or their children need special therapeutic assistance.

The basic philosophy of parent group education is perhaps best expressed in the following statement by Dr. Peter Neubauer:[2]

"*Education* is aimed at those faculties of the ego which are undisturbed by conflict. It is oriented toward the healthy factors of the personality, and appeals to the ability to judge, to gain understanding, to learn to use one's experience for new and different situations, to plan, to make choices, to adapt to changing circumstance, to add new experiences.

"The term education, in connection with group experiences, is not used in the traditional sense of applying only to the intellectual capacities of the group members. Here the educational experience takes on a broader meaning. It recognizes the importance of feelings and attitudes and uses emotional mobilization as well as intellectual stimulation. It uses all the potent psychic factors of the educational process, while maintaining an awareness of the difference between education and therapy.

"And while in group education, there is an awareness of individual problems and their effect on the parents' functioning, the unconscious motivations of emotional problems are not explored nor are attempts made to resolve these problems. The purpose of the group is not to discover, expose or heighten their sense of their problems, but rather to explore, develop and reinforce the health they have."

In descriptions of community agency programs, published papers of professional conferences and indices of periodicals, there is a noticeable dearth of reference to group educational programs for married people which focus specifically on the usual areas of concern in marriage. Are there areas in marriage where education *can* serve to prevent difficulties and support the positive aspects of adjustment? The foregoing description of parent group education may provide several clues as to the possibilities of group education in marriage.

Where the difficulties in a marriage are due to intra and interpersonal conflicts, there seems little question that a therapeutic effort is needed if the problems are to be resolved. However, the fact that a marital partner has neurotic difficulties will not neces sarily create problems in the marriage.[1] In such instances, a group educational experience can be helpful in strengthening the marital relationship.

There are areas, particularly in a new marriage, where difficulties may be due to lack of knowledge and experience. Such lacks may also aggravate those problems which are neurotic in character. Many newlyweds have had little or no experience in practical matters, such as budgeting and household management. More important, though, are the areas of relationship which are new and at times difficult. Important among these are the reality of the marriage as compared with the premarital fantasy, getting to know another adult's special characteristics as they are revealed in daily intimate living, the altering of each mate's pattern of daily living in relation to the needs of the other as well as of himself.

Some of the facts about these aspects of marriage can be obtained from the literature. But what many people need, along with the facts, is the opportunity, through a shared experience with other couples in similar situations, to see all aspects of what is involved in these problems. Among other things, they need help in understanding how their feelings about money, about changes in the nature of their personal freedom, about family planning, can affect their handling of these facets of their marriage.

The educational groups for engaged couples have anticipatory guidance as their primary purpose, as do educational groups for expectant parents, particularly those about to have their first child. In the latter groups, couples not only discuss their current experiences and attitudes in regard to the pregnancy, but think through what they want to do about such practical matters as the husband's presence during labor, about rooming-in, and breast or bottle feeding. Also, they can plan ahead for the first weeks with the new baby.

As a marriage goes on, other situations arise where participation in an educational discussion group may help prevent trouble. A problem in many marriages is that of having to combine households in order to take care of aging parents. A joint living arrangement may create tensions and difficulties even for couples who have had a basically good relationship with their parents. Some of these are often related to lack of knowledge about the special problems of the aged. An educational approach which includes discussion of the emotional aspect of the experience may well prevent serious problems from developing.

Although the husband-wife relationship has unique aspects, many of its areas overlap those of the relationship of the engaged couple, or of the man and wife as father and mother. Experience in premarital and parent education groups suggests that extending the work in group education, under skilled leadership, to focus on areas of marital relations, with the goal of preventing difficulties and supporting the strengths in marriage would be worthwhile.

References

1. Ackerman, Nathan, "The Diagnosis of Neurotic Marital Interaction." *Social Casework,* XXXV, No. 4, 141, 1954.

2. Neubauer, Peter, "The Technique of Parent Group Education: Some Basic Concepts." *Parent Group Education and Leadership Training,* p. 11. New York, Child Study Assn. of America, 1953.

II Group Counseling
SANFORD N. SHERMAN*

THE TERM "group counseling," as used here, refers to a specific and delimited form of psychological treatment, as distinguished from educational or didactic guidance methods to which it is sometimes applied in general usage. Group counseling may be compared to casework counseling. Both are based on a psycho-social diagnostic evaluation of the individual, and both have similar treatment goals for the individual which are sociopsychological in character.†

Group counseling is a treatment method most readily employed in social agencies, where the applicant usually stresses the social, interpersonal aspect of his problem—his poor marital relations.[3, 4] Its central aim, as that of family casework, is to improve or restore the family balance and the social functioning of the individuals as marital partners. This requires that out of a diagnostic evaluation of the "whole," which includes the individual and the family of which he or she is a part, there must be a sectioning off, a delimited area toward which the therapeutic effort is to be directed. The areas of the individual personality which receive therapeutic attention are centered in the ego —in conscious and preconscious attitudes which contribute to the imbalance of the marital relationship.

In some respects this orientation may be likened to the Clinical Sector Psychotherapy, recently described by Deutsch and Murphy, in that "Only a sector of the personality or one problem will be treated, whereas other parts of the patient's personality

---

* Assistant Executive Director, Jewish Family Service, New York.
† See Chapters XIV and XV.

and other problems remain untouched." [1] Thus, group counseling will define as the area for treatment a troublesome aspect of a marital relationship, and aim to bring about changes in this relationship and in the functioning of either or both clients as husband and wife.

Although this procedure is similar to psychotherapeutic sectoring, and its orientation is that of dynamic psychology, it differs in its emphasis on the social dimension. Regarding psychotherapy, Deutsch and Murphy state: "In a scientific approach the change in the inner psychic dynamics should be the chief determinant for the evaluation of the results." [2] However, in group counseling's evaluation of results the weight of emphasis is the change in social relationship and behavior, whether or not accompanied by an identifiable change in inner psychic dynamics. [6]

When a marriage problem is posed by an applicant, we recognize that the conflict situation almost always has a social and personal interrelationship and inner psychological elements, and that certain aspects of the personality of each marital partner are contributory and interactive. We are therefore concerned with a triad of factors: the behavior, psychological needs and defenses of each of two individuals and their interaction in the marriage. Hence there will usually be two clients. The therapeutic concern will not only be centered on each individual treatment process, but also on how each step in the treatment of both partners infringes on and influences the marriage.

The marital relationship is further sharpened as an area of therapeutic preoccupation when the other marital partner is also engaged in treatment. Both ends of the axis of marital relationship are now involved in a corrective endeavor. If perceived and correctly utilized by the worker or workers, interaction between the two treatment processes leads to mutual enhancement of the relationship.

Indications for group counseling as the treatment of choice can be negative or positive. Group counseling seems indicated.

for the patient who shows a persistent tendency to deny the psychological aspects and to overemphasize the social aspects or the behavior of the other marital partner in the marriage conflict. Sometimes, too, a particularly slow movement in the individual treatment of a client with a character disorder, whose defenses are strongly mobilized against involvement in an individual relationship may indicate such an alternative approach.

Group counseling is also indicated where the individual's social intercourse is unusually constricted and where the social interaction features of the group can contribute to ego building. These are a few of the many considerations which point to a differential utilization of group counseling. Some of them parallel those of group psychotherapy,[5, 7] some are already known empirically, and many have yet to be identified through further experience and research.

The group to which the marital partner is introduced may be composed of six or seven other individuals whose presenting problems are either marital in character or mixed. The group may be all of one sex or both. In no instance are two individuals from the same family included in the same group. In our experience, the person in a group with both sexes will have a wider range of relationships within that group.

Transference of attitudes and patterns of behavior expressed characteristically in marital life are commonplace. A group composed exclusively of women will discuss sexual relationships or competitiveness with men more quickly and easily than they would in a mixed group. But in the mixed group there is the involvement in relationships with men as well as women in a therapeutically geared situation. Projections and distortions become more quickly apparent to the group as a whole and the individuals in question, and newer, less disordered attitudes find a place in their adult egos. Directed and restrained by the group counselor, this emotional interpersonal experience is confined to verbalization rather than allowed to be acted out.

The group counselor treads an obstacle course lined with

pitfalls. His aim is clarification of attitudes, reduction of emotional conflict, diminution of anxiety and, often, strengthening of defenses or their modification. Far from covering the full range of personality function of the individual, his goal is concerned only with those factors bearing on the relationship balance between the marital partners.

The counselor avoids didacticism, unless it is indicated at particular points in order to meet the emotional need of the client. On the other hand, he also avoids broadening the range of treatment effort beyond the delineated sector of personality or intensifying the treatment probe beyond conscious or preconscious areas of ego function. Transference, especially negative transference, needs to be recognized, understood and dealt with, but not in the direction of deepening the transference neurosis. Usually, attitudes, hostile or positive, are handled in a counseling group, rather than the underlying transference. The counselor tries to establish and maintain the proper kind of atmosphere for group discussion by making specific contributions or responses in such a way that the social, marital and familial situation remains in the forefront of psychological discussion.

The following is an abstract from the record of a typical group counseling session:

In a discussion by a group of five women and two men, Mrs. A. had asked the counselor a direct question which he answered briefly. Mr. T. said he did not mean to be critical but, though he had understood the group counselor, he was sure Mrs. A. hadn't and asked that the counselor try explaining further.

Mr. T.'s assumption that Mrs. A. would need a fairly simple point explained again was related to his frequent transfer of attitude from his wife to Mrs. A. For some time, Mr. T. had been expressing veiled hostility toward Mrs. A. in whom he saw a replica of his wife's sexual constrictiveness and emotional demands. Of course Mr. T. was also expressing hostile feelings toward the group counselor by his comment.

The counselor asked Mr. T. what he really felt and wanted at that point. Mr. T. became confused and said he was no longer certain. Maybe he was again feeling his own inadequacy in communicating with and "getting across" to his wife.*

Two other women in the group, as well as Mrs. A., each with some heat, said they had understood the counselor and that Mr. T. was being critical out of his own need. Mrs. A. said she shriveled now as she always did when Mr. T. criticized the counselor, and she had no doubt he had intended criticism by his very opening comment. She and the two others were sure Mr. T. was wrong. The counselor had been very clear and they had understood very well what was said.

Mrs. B. felt it was a question of resistance to grasping a comment which touched one's emotions. In the varying comments made by the three women, there was a defense of the counselor and an attack on Mr. T. (For some time past Mr. T. had frequently touched off hostile responses from Mrs. A. and Mrs. B. because to them he personified the essential passivity and lack of responsibility they ascribed to their own husbands.) The counselor encouraged further discussion of the women's distress.

Mrs. A. expanded on her feelings whenever Mr. T. seemed at all critical of the counselor. It reminded her of her chronic feeling that her husband was a fake who pretended to know more about everything than anybody else. The counselor offered the suggestion that Mrs. A.'s husband challenged the world's wisdom and Mr. T. challenged the counselor's. Mrs. A. quickly associated this to her feeling that her husband always fell far short of the standards set by her father. In her mind her husband always came out second best in comparison with her father, and maybe she ought to stop making such comparisons.

This reminded Mrs. B., who had lost her father in her early childhood and who had an infantile dependency on the counselor, that she had also responded to Mr. T. as she did constantly toward her husband, always comparing the latter unfavorably with a romantic ideal she knew could never exist. Mrs. B. recognized that a triangular situation existed for her at this moment in the group, involving herself, Mr. T. and the counselor. She was hostile to Mr. T. and competitive with him because he was not like the counselor.

With deepened understanding, she now saw this to be a repetition of the triangle she created at home: herself, her husband in-

---

* Mr. T. was an active part of the ensuing discussion, but in the interest of keeping the focus in this abstract on Mrs. A. and Mrs. B., his activity is not being reported.

stead of Mr. T., and a fantasied third person instead of the counselor. Who, she asked, was this non-existent third party with whom she always contrasted her husband unfavorably? She realized this was an area of her emotional life that needed further clarification.

The remainder of this group session was devoted to a discussion by the clients on the subject of transferred attitudes toward spouses. The session had some elements of clarification for a number of persons in the group. For Mrs. B. in particular, it was a point of precipitation of self-understanding and a progressive step in ego definition. Her long-standing confused dissatisfactions and resentments toward her husband, which were projective and attended with anxiety, were on the way to clarification. The interaction in the group and the many relationships it afforded her (identification of herself with Mrs. A.; her husband with Mr. T.; and the all-good father ideal with the counselor) provided therapeutic dynamics which can only be found in part in the individual treatment situation.

Unlike group psychotherapy, the purpose here was not the further analysis or intensification of the transference phenomena occurring in the group. Rather, the influence of transference on the relationships in the group was *recognized* by the counselor, while he *highlighted* the conscious relationships and attitudes in the group and in the outside marital relations. As has been noted, casework and counseling confine their therapeutic efforts to the adaptive functions of the ego. These methods delineate partialized socio-psychological areas with which to work, and define their goals in terms of improvement in social function and marital, parental, adult-child relationships.

References

1. Deutsch, Felix and Murphy, William F., *The Clinical Interview*, Vol. II, p. 14. New York, International Universities Press., Inc., 1955.
2. Ibid., p. 15.
3. Grunwald, Hanna, "Group Counseling in a Casework Agency." *International Journal of Group Psychotherapy*, iv, No. 2, 183-92, 1954.
4. Hulse, Wilfred C., "The Role of Group Therapy in Preventive Psychiatry." *Mental Hygiene*, XXXVI, No. 4, 531-47, 1952.

5. ———, "Dynamics and Technique of Group Psychotherapy in Private Practice." *International Journal of Group Psychotherapy,* IV, No. 1, 1954.

6. Sherman, S. N., "Casework-Oriented Group Treatment in a Family Agency." *Casework Papers of 1955.* New York, Family Service Association of America, 1955.

7. Slavson, S. R., "Criteria for Selection and Rejection of Patients for Various Types of Group Psychotherapy." *International Journal of Group Psychotherapy,* V, No. 1, 1955.

III   Group Psychotherapy

WILFRED C. HULSE, M.D.*

PRESENT UNDERSTANDING of group phenomena is based to a large extent on a scientific appraisal of the primitive roots of group psychology. The attempt to influence social and moral attitudes through a group approach may be traced back to the beginnings of human society, and knowledge of these natural group processes has influenced the development of professional techniques in prevention and therapy. Our goals are the stabilization and reorganization of ailing marriages, wherever possible, or the prevention of further damage to individuals and society from a marriage which lacks sufficient positive elements to justify its existence.

There are three distinct approaches to group treatment of marital problems: group education, group counseling and group psychotherapy. Group education and group counseling are focused on the marital problem itself, though the methods and goals of each are essentially different. Group psychotherapy is specifically focused on the psychopathology of the individual with marital difficulties. The primary purpose of group education is the prevention of marital breakdown; that of group counseling the modification of disturbed relationships in marriage; that of group psychotherapy the cure or improvement of the underlying emotional illness in one or both marital partners.

Group therapy establishes a multi-phasic interplay of transference phenomena and other subconscious emotional reactions, which can be used therapeutically by a well-trained therapist, but which in their intimate interrelationships are still largely un-

---

* Associate Attending Psychiatrist, The Mount Sinai Hospital, New York; Chairman, International Committee on Group Psychotherapy.

known and subject to further research. Its origins may be traced back to two distinct and separate roots: One is the psychological processes of the primitive group and their early application to social controls and a kind of aboriginal mental and social hygiene. The other is the development of individual psychotherapy and psychoanalysis during the last fifty years.

Psychotherapists have attempted to use the group directly for the treatment of social ills and for the reform of social maladjustments, such as treatment of the sick marriage as a social phenomenon. Trigant Burrow's efforts in this direction through "phyloanalysis" have never found an appreciable following or wider application.[1] Psychodrama with the use of an "auxiliary ego" representing the marital partner and trying to relieve marital tension by acting out in the dramatic therapeutic situation has also found only a limited area of application.

An analytic attempt to deal with both marital partners therapeutically in a "group of three" (husband, wife and analyst) ended inconclusively due to the untimely death of the New York psychoanalyst, Sam Rosen, who told this author that he planned to publish his experiences after the successful conclusion of fifty cases; he had treated only thirty couples at the time of his death.

The psychoanalytic treatment of both marital partners in the same group has been tried by a number of group therapists,* but their experiences have not led to conclusive publishable results. According to some reports, the treatment of marital partners in the same group focused on therapy for psychosomatic illness has provoked explosive and often dramatic reactions.

It seems to be generally accepted in group psychotherapy, however, that members of the group do not profit from therapy if they share the therapeutic group with family members, and that in such situations the danger of acting out transference into life situations increases. Many experienced group psychothera-

---

* Alexander Wolff, W. C. Hulse and others.

pists discourage even social relationships between members of their groups. Such precautionary measures, based on theoretical considerations and practical experiences, seem to mediate against *direct* psychotherapeutic group approaches toward marital conflicts.

The parallel treatment of the marital partners in two different groups conducted by the same therapist or by different therapists in the same clinic-setting apparently avoids some of these pitfalls. Such groups have been conducted, but they lend themselves to counseling rather than to psychotherapy. This is due to the tendencies of homogeneously selected patients to stick to the external purpose for which they have joined the group (in this instance marital problems), rather than permit themselves to open up their deep individual conflicts and expose them to the therapeutic process.

This phenomenon poses an important question regarding the therapeutic assignment of patients seeking treatment as a result of marital difficulties. Patients may try to avoid entering the area of their marital incompatibility for extended periods of time in an individual psychoanalytic or psychotherapeutic process. In group psychotherapy, discussion of actual marital problems cannot be put off due to the more active interference by other group members. As "auxiliary therapists," they steadily confront the patient with the marital, extra- or premarital problems of the group members.

The psychotherapeutic group process, on the other hand, is open to another kind of resistance in this type of patient. By dealing primarily with the social aspects of a poor marriage, the patients can aid each other in resisting the exposure of the deeper dynamic conflicts which are at the bottom of their marital difficulties, as for instance latent homosexuality and other severe psychosexual pathology.

It is therefore of great importance to carefully select those who might profit from group psychotherapy, rather than from other psychoanalytic or psychotherapeutic methods. The marital adjustment of a specific patient can be dealt with from different angles in group and in individual therapy, the former approaching such problems often on a more reality oriented and less isolated level than the psychoanalytically oriented individual process.

Most group psychotherapists treat their patients once or twice a week for 60 to 120 minutes. There are usually six to nine patients in psychoanalytically-oriented interview groups. Heterosexual as well as monosexual groups are used. Most groups are "open," i.e. patients leave the group when they have achieved the optimum of therapy, or if further treatment in a group seems inadvisable. They are then replaced by new patients to keep the group at a full level of participation. The technical aspects of a therapeutic group dealing with marital problems are not specifically different from other group psychotherapeutic settings.[4]

The impact of successful group psychotherapy on the marriages of neurotic and ambulatory psychotic patients is strong. Psychotherapy has as its goal the modification of symptoms and the better adjustment to life situations on a permanent basis. Such goals can often be achieved only through character changes. While it is a matter of controversy as to whether character changes can be achieved by any therapeutic method short of prolonged psychoanalysis, great modification of character and personality traits have been observed and reported by researchers and practitioners in the field of group psychotherapy.

Where such changes occur and contribute to a better and more satisfactory marital relationship, they very often also establish the need for direct help for the other marital partner, who may be unprepared or unable to deal with the changed attitudes of the recovering patient in therapy. Casework or psychotherapy may be necessary to help the untreated marital partner in adjusting to marital life with the recovering patient.

On the other hand, the untreated marital partner may very often profit greatly from the improvement of the patient, by gaining insight into some of his own maladjustments (if he is not very ill himself) due to the improved behavior of his mate and a certain psychotherapeutic help applied to him "by proxy" by the recovering patient.

Group psychotherapy, through specific therapeutic phenomena, can exert very specific influences on the marital situations during the therapeutic process. This method permits the patient to compare his own fantasy life, in the presence of the therapist and the adjunct therapists represented by the sibling-group members, with the life situations of other patients as well as with their psychodynamic difficulties. Patients are able to recognize identification and de-identification phenomena in themselves and in others.

The fact that they project their infantile wishes to be nurtured by their mothers upon their husbands or wives is frequently recognized by patients as occurring in other group members. They gain a kind of pseudo-insight into the dynamics of other group members long before they can accept similar dynamics in themselves. Insight might then occur as "a flash" or as a result of the therapist's interpretation. But it is achieved more often by the active therapeutic function of other patients whose explanations or interpretations are frequently accepted with less resistance and greater potentiality for permanent integration than the active interpretations of the "professional" therapist. The patient's need to become a therapist himself finds more gratification in the group than in the individual setting. This process also has an ego strengthening effect closely associated with multiple transference phenomena.

During states of improvement, marital difficulties diminish and often disappear. The patient becomes less anxious and irritable, and is willing to take on greater responsibilities. Libidinal forces, formerly bound by anxiety and narcissistic preoccupation, are freed, stimulating sexual activity and promoting sexual satisfaction. Patients may try to apply the greater freedom of libidinal attachment at first within the group, trying to act out toward other group members or toward the therapist. Adequate timing of interpretations will achieve actual adjustment in the life situation, rather than permit a pseudo-adjustment in the therapeutic group. It is necessary to watch the reaction of the marital partner closely, and through counseling help him to meet the emotional needs of the changing patient.

While many psychoneurotic and pre-psychotic patients can profit from group psychotherapy so long as they are not hallucinating or severely depressed or mentally deficient, the specific effectiveness of this method seems to apply to patients with very rigid super-egos—many types of character disorders with compulsive, obsessive and phobic symptomatology, as well as hysterical character disorders. It is, of course, not a coincidence that such personalities also have frequently severe and long-standing marital conflicts, and that they are very difficult patients in individual psychotherapy. A certain number of chronic anxiety states are also responding well to group psychotherapy, but they usually need individual sessions between group sessions, a combined method that has many advantages.[3]

Group psychotherapy in patients with marital problems permits more catharsis and self-participation, greater reality closeness and super-ego relaxation than most of the other psychotherapeutic methods. In addition, the group provides certain re-educational features not easily obtained in an individual therapeutic setting.

The importance of reality testing and working-through in the group therapeutic process has been emphasized before.[5] Its value is seen particularly in advanced patients who have gone through an extended analytic process and are then able to experience and integrate reality if it occurs in a supportive and anxiety-reducing therapeutic situation.

The following brief excerpt from a case history illustrates the actual group process and its effect on the marital situation:

Rose, a commercial artist of Irish descent in her twenties, whose early case history has been reported elsewhere,[2] had returned to group psychotherapy two years prior to the situation described here. An anxious and dependent character, she had never received adequate love and support in her childhood. In her immature search for warmth and acceptance she had been seduced and impregnated in adolescence. Later, when her resentment against her emotionally and socially limited husband had led her into a variety of extramarital adventures, she returned to group therapy.

Increased insight and ego growth had developed slowly during the last two years in therapy, but her ambivalence toward her husband and her need to seek acceptance and reassurance through extramarital affairs had not diminished. The group members (all women) did not openly disapprove of her marital delinquency. Instead, they had frequently shown admiration, envy and hostility in the face of her daring and unconventional search for sexual gratification, even if her search for warmth, love and a sense of belonging was unsuccessful.

One evening after a group session Rose's husband picked her up in his car and gave several other group members a lift home. In the next group session, there was much talk of Rose's husband, of his gentleness, his good looks and other desirable qualities as a mate and companion. From then on the members of the group continued to probe into his social life and also into the intimate details of his sexual performance. He became a highly libidinized object of their fantasies. The emotionally disturbed personalities of Rose's two extramarital lovers were analyzed. Rose's supposedly good sexual adjustment was unmasked as a distorted identification of sex with a delinquent act that could only be enjoyed in an extramarital rather than legitimate setting.

The group members, siding in amorous fantasies of their own with Rose's husband, proposed that he would become sexually more aggressive if Rose could control her own aggression and accept a more feminine and submissive role. The process of working-through of this conflict continued over the better part of a year. During this time, Rose's sexual and intra-family relationship with her husband improved greatly. She gave up both lovers. She now has a satisfactory marriage and is a more mature woman with many new achievements in her social adjustment and artistic career.

In conclusion, the selection of patients for group psychotherapy should be based on the specific needs of the patient with marital conflicts. The argument that group psychotherapy or all group methods are applicable when the scarcity of trained personnel or the patient's financial status does not permit the individual approach is not valid. Group methods are not inexpensive, nor are they a short-cut in therapy.

Group psychotherapy is a slow process. Its effects on the patient and on his marital situation do not become manifest until considerable skill, time and effort have been invested. Group therapists or counselors need extensive experience in individual psychotherapy or casework as well as specialized training before they can apply themselves to working with groups.

References

1. Burrow, Trigant, *The Neurosis of Man.* New York, Philosophical Library, 1953.

2. Hulse, Wilfred C., "Symbolic Painting in Psychotherapy." *American Journal of Psychotherapy,* III, No. 4, 1949.

3. ———, "Transference, Catharsis, Insight and Reality Testing During Concomitant Individual and Group Psychotherapy." *International Journal of Group Psychotherapy,* V, No. 1, 1955.

4. Slavson, S. R., Ed., *The Fields of Group Psychotherapy.* New York, International Universities Press, 1955.

5. ———, *An Introduction to Group Therapy.* New York. The Commonwealth Fund, 1949.

CHAPTER XVII

Problems in the Prediction of Marital Adjustment

NORMAN REIDER, M.D.*

THE VERY FACT that in the history of the human race the importance of the family as a primary social unit has almost never been questioned seemingly commits us to make every possible effort to maintain the family and remove marital discord. Yet these efforts, however worthy their purpose and aim, have not escaped all prejudicial factors; their underlying premises require careful examination.

By prejudicial I mean those irrational elements, highly emotionally tinged, that make us automatically consider possible divorce or separation with a certain amount of discomfort and anxiety. Sometimes this attitude reflects the patient's feelings; sometimes it represents conflicts which are stirred up in ourselves, over and above a rational consideration of the possible damage to result from the disruption of a marriage.

* Chief, Psychiatric Service, Mount Zion Hospital, San Francisco.

True, damage may result from one or the other or both partners, or for the children; yet it has been my observation that even though sound clinical judgment indicates that a couple will both be better off if they are divorced or separated and that even the children may be better off, elements of the irrational unconscious, conscience, or fear of what relatives, friends, neighbors and society in general may think, help to obscure the reasonable approaches. Of course, such matters as shame and embarrassment, whether justified or not, must not be ignored. They must be dealt with as any other data, realistic or neurotic—by investigation, evaluation and good clinical judgment.

Frequently, the very orientation of a study of marital problems has, whether we like it or not, a certain prejudicial character in that the therapist or counselor is committed not to investigating the individual and his whole situation, but to preserving the institution of marriage. Consider how much more catholic and comprehensive its approach if the thesis also included studies on what the definite indications for the disruption of a marriage are. To continue some marriages, given certain insoluble conditions, may be a disservice to both the marital partners and to the children as well. All too often the disruption of a marriage comes as a final admission of failure and as a last resort, rather than as the result of scientific efforts at study of the total situation. Perhaps such a traditional attitude is the proper one, however, since a calm, scientific approach to marital problems with an equal readiness for divorce as well as for correction is most difficult, and the method would lend itself to all sorts of abuse in the hands of those who capitalize on the misfortunes of unhappily married persons. The method would also require a considerable amount of courage in this age of anxiety wherein the quest for security becomes more and more ominously palpable. We live in an era wherein tightening and fortifying of bonds and ties is the order of the day rather than any lessening of them. Nevertheless, studies of human relationships should be used to benefit people and not institutions. If, for example, it is true that a marriage between a schizophrenic and a manic-depressive usually does not work out, as Dr. Jacobson has suggested,* then this is a fact worth knowing, and deserves to be dealt with in some detail.

As practitioners and clinicians, we are particularly oriented to the recognition of the disastrous effects of separation and divorce; they are constantly before us. The many contradictions inherent in common concepts as to the catastrophic effects of separation and divorce have recently been dealt with in detail by Dr. William J. Goode.[1] Moreover, the several references in this volume with regard to the tendency for an untreated neurotic to make a failure of a second marriage, a fact we substantiate from our practices, is not corroborated in Dr. Goode's study. On the contrary, Goode points out that nine out of ten divorcees who remarry *claim* their second marriage is a better one. This sociological finding is a more accurate appraisal than one derived from our clinical experience, although the interpretations of the differences are of no small importance. Thus, our limited attitudes may keep us from using important data which is not derived from our own work; this is particularly true with regard to Kinsey's findings.

The premise that we are somehow or other, and for good reasons, largely committed to the preservation of marriage and to making it a happier institution includes other kinds of prejudices, some apparent and others implied. Although clearly, the character of marriage has no homogeneity, the trend to treat it as an entity persists; a good marriage is considered a rather uniform phenomenon, composed of similar ingredients in each instance. But this is not true. However many social data are gathered, however many diagnostic typologies are classified, the phenomenon known as marriage in our society has no homogeneity. Instances of marriages arranged and settled by the parents of the couple still exist. Multiple and complex motivations,

---

*Chapter VIII.

conscious and unconscious, exist in every marriage—the quest for security, for a repetition of a previous home life, for convenience, for changes in social status, for sexual activity in a socially approved setting, for children, for proofs of maternity or paternity, for satisfactions and needs of various kinds, including those of punishment.

Even such complex, diverse motivations in marriage would still make for a relatively simple situation if they always remained the same, but the problem is further complicated by the fact that people change and their needs change with them as they grow older. The marriage that is sound or seemingly so at its beginning because the partners have certain needs in common, may become difficult as changes occur. These changes come about not only because of disappointments or the lessening of early passion. The vicissitudes of the libidinal and interactional changes that occur in marriage introduce another kind of complexity. Many a person matures in the course of marriage and if his partner meets these qualities of maturation, things are likely to go well, though not necessarily, because the process of maturation in one person may then uncover needs which differ greatly from those of his partner. Perhaps the best example of this type of phenomenon is what occurs when two people marry who are not in love and who do not fall in love with each other as time goes on, although usually such a marriage involves some type of neurotic living out of an unconscious fantasy. Such things happen, and I cite them mainly to show the many complexities that surround marital problems. They appear in any cross-sectional analytic view and in the experienced clinical examination of the life histories of many marriages.

Considered, then, from such points of view, any set of rules or injunctions seems unlikely to serve as a worthwhile guide to successful marriage. And yet a considerable body of empirical experience qualifies this statement. From time immemorial, whenever there has been any semblance or illusion of freedom in choice of a marriage partner, people have used rules, signs and tokens of one sort or another that have worked. Being in love, coming from a certain social group, bearing some particular physical characteristics, having a similar or different temperament, or being given magical signs by a fortune-teller—all have been so used. Many a marriage has withstood the pragmatic test, so that people believe the omen to be of value.

In our scientific efforts we tend to lose sight of the considerable amount of magic, let alone wishful thinking, that goes into preparations for marriage. The mores and ceremonials of courtship and betrothal, the looking for signs of auspicious days, the vows and rituals, the many and varied incantations for best wishes and good fortune, the hopes, the prayers, the gaiety and tears that attend a marriage are mainly devices attempting to insure happiness in marriage.

Of course, there is more than all this in whether a marriage succeeds or fails, and whether the magical rites work well or not. Perhaps the old rites play less of a role in our present day and age than acceptance of traditional ceremonies. But the ceremony still symbolizes the entering of a pact or contract that is to be taken seriously; and indeed the ceremony itself and all the atmosphere of seriousness, fortified at times by legal and religious restrictions, help to insure, for most people, a certain degree of stability and permanence.

Our primary concern, though, extends further than permanence. We seek to know whether actual contentment and satisfactions exist in a more mature sense. For the examination of these more meaningful features we have a body of psychiatric and psychoanalytic knowledge that provides numerous sets of criteria for application to the problems.

Résumé of Studies

How to use Kinsey's data or Goode's studies is another matter. Perhaps at present we can best use them to check our clinical judgments. How to use our own clinical material is still a greater problem in a sense. In the course of our work we look for cues and constellations of data that we may use in a predictive way. We learn from our experiences by making hypotheses constantly, whether we realize it or not, and in doing so, we are constantly testing our predictive capacities. And we try to make this part of our work as substantial and yet as flexible as possible, so that we can communicate our findings to others. We correct our impressions, re-orient ourselves, re-explain our findings and form newer ideas. One aim, again whether we realize it or not, is to concentrate our skills and knowledge into communicable short-cuts and thereby to help our patients and clients. It seems then in order to examine our present knowledge in an effort to apply what we know in a predictive way. Do we really know enough to say that this or that marriage will work or fail?

It would be an interesting topic for research to collect and study the predictive pronouncements of friends and relatives in regard to a marriage. Although such predictions are not based upon any scientific aptitude, they have some accumulation of clinical intuition, sometimes called wisdom, which might form the basis of a worthwhile study of a certain kind of homey, commonplace apperception of character traits and prediction of future course. "Mary is a good girl, she'll make a good wife." "John has sowed his wild oats and he's going to settle down now and be a good husband." "Jane is going to make this marriage work. She's not going to make any of the mistakes her mother made." "Tom worships her, and he's the sort of fellow who will never see any of her faults. So long as she gets what she wants, everything will be all right." In order to systematize such highly individual prejudices and residues of vague experiences, a study of this kind might be only a small part of a larger one on the formation of opinion. Otherwise, we take little stock in such opinions; they are unsystematized and so we consider them unscientific. Because we are bound by nature to systematize our knowledge we are led, whether we realize it or not, to the establishment of various orders of typologies.

The first of these typologies is the strictly diagnostic one. A quick examination of our usual psychiatric nosology shows it to be relatively meaningless in this regard. If we keep in mind the degree of subjectivity in the designation of a good marriage and what constitutes a good marriage, we can see the uselessness of trying to establish any criteria by the study of diagnostic categories. Two schizophrenics who marry may get along miserably. Others marry and often get along quite well. Reviewing both my clinical experience and my observations among people of my acquaintance, I have to conclude that I've seen "mixed marriages" between one diagnostic category and another both succeed and fail. Nor am I impressed with the possible predictive value in such diagnoses, with the one possible exception that I have known few marriages between male paranoid characters and female hysterical characters to succeed, by almost any criterion of success. The nature of the sado-masochistic struggle in these settings is of an unbearable quality. From a purely nosological attitude, therefore, one may as well attempt to find correlations of the success in marriage between the color of men's eyes with the color of women's hair.

Nor does the typology arising from considerations of character structure provide data more amenable to our investigation. With the exception stated above, I have not seen any particular value in this approach, whether genetic or descriptive criteria are used. I know of no studies from the point of view of whether the partners got along better if they were both oral characters, or got along poorly if one was an oral character and the other an anal character. What might at first appearance seem to be more valuable, descriptions of character types such as "passive-dependent" or "aggressive," also fail, on closer examination, to provide any worthwhile working hypothesis. Clearly the phenomenon to be studied is too complicated for any base provided by diagnostic classifications. Other considerations, therefore, must be looked into.

The emphasis that has been placed and will continue to be placed on scrutinizing the predictive value of various sorts of investigation has several implications. First, a similar kind of technical difficulty confronts us here, as in other phases of psychiatric work—the problem of whether, having arrived at some sort of diagnostic formulation, we are justified in turning it into a predictive tool. Much of the material in the studies reported in this book has already proved most valuable from a diagnostic point of view. Especially valuable trends are seen in the papers that make a "family diagnosis," a short-cut expression for conflict arising from antagonistic drives not only in the individual but in the family group. Such diagnostic formulations have demonstrable value as the basis for a plan of treatment.

Again, although I would not dispense with all personality tests or diagnostic criteria in assessing marital successes and failures, we must recognize that the roots of any test or predictability scale, however scientific the design, have not yet been perfected. It is understandable that one should wish to know the future, to find out whether or not taking a certain important step will bring happiness and contentment. But we must consider how far predictive psychological techniques have advanced to this goal.[2]

A review of the studies on the use of psychological tests is now in order, for these tests may give clues as to the qualities of interaction between marital partners. For instance, Harrower's point, that the least disturbed couple often comes to treatment first, is certainly significant.* However, she interprets her test re-

---

* Chapter X.

sults as indicating a good prognosis when the marital couple has parallel scores at different levels, or similar distortions of thinking; while excessive and deviant scatter indicates a doubtful prognosis.

Harrower takes the position that something in each pairing of tests points to a good prognosis, with indications of a good outcome from therapy. This in itself, though seemingly cautiously and correctly stated, does not give enough clue as to what interaction makes for the difficulties. One also gets the impression, a frequent one in test situations, that factors derived from the test have been oversimplified. The impression is especially clear in conflictual situations which we know, clinically, are overdetermined. The test work often implies, too, that what is etiologically responsible for the marital difficulty is something within the test pattern. The mere finding of differences in tests should not in itself lead to the assumption of any etiological factors.

Involving even greater danger than does the methodological approach is the possible use of the test techniques for screening and prediction without the benefit of the use of clinical material. The same criticism may apply to some of the conclusions reached in Piotrowski's contributions on the use of M.* In his most interesting study of marital compatibility, there is still the implication that M is valid and stable proof of compatibility or incompatibility, rather than one mode of approach to the problem. Again we need a word of caution about the use of tests as predictive devices at this stage of our knowledge.[2]

Surely the most illuminating studies come from the excellent psychiatric and psychoanalytic presentations which give in detail the nature of the interaction of conflictual forces within couples. In my opinion the most valuable are those which indicate the kinds of adaptation that occur in spontaneous adjustments in a successful marriage or those adaptations that occur as a result of treatment. Of course, such clinical studies also point to sources

---

* Chapter XI.

of difficulty and give clues as to their use in a predictive way. But caution is again emphasized, and justly so, since the study of marital relations should in no way differ from other areas of psychiatric study in which prediction out of past clinical behavior is too hazardous under the multiple conditions to which the life of an individual is subjected.

Prospects of Improvement of Marital Relations via Psychiatric Techniques

One conclusion, at the present state of our research, would be to the effect that many a marriage could be saved if one or the other or both partners went into psychiatric treatment, and such a conclusion would be warranted on the basis of experience. It implies that treatment would be the primary way of meeting the problem. Obviously, this is not always feasible. Therefore, it is justifiable to anticipate some sort of prophylactic devices for meeting problems before they arise, and this thought is back of all kinds of planning toward making marriages more effective. Suggestions and advice range from trial marriages to giving courses in domestic science, and encouraging children to "play house." The theory back of this is clear. By way of anticipation, there may be some working through a situation and thus an actual preparation for mastery of a future area of difficulty. This leads to a consideration of the whole area of education, which ranges from courses in family life in high schools and colleges to premarital counseling, and the efficacy of the use of marriage manuals.

The latter subject has been treated very adequately by Sapirstein, who so aptly points out that many individuals use the material in the marriage manuals as measuring rods and construct artificial levels of aspirations to which they feel they have to conform, especially in regard to sexual activity and behavior.[3]

Yet despite the artificialities that may result from certain types of conformities in sexual behavior, we frequently overlook a positive gain that may come from the so-called educative devices. We are usually so accustomed ourselves to the use of conceptual approaches to the problem on a highly sophisticated level that we lose sight of the fact that many people do have a great deal of curiosity and quest for definite types of information. We overlook the fact that many people can use simple knowledge about anatomy and physiology as an aid to master, in some belated sort of way, troublesome problems. To be sure it is no accident that courses and books in marital relations often provoke anxiety instead of allaying it; but, again, whether they produce adverse effects is frequently influenced by an already existing tendency in the individual toward anxiety. It is these failures in the popularization of knowledge that we come across as clinicians; and because the beneficial effects of popular education rarely come to our attention, we tend to forget them.

This trend toward popular education by courses in marital relationships in high schools and colleges and by marriage manuals is bound to increase as part of our general trend to give as much information as possible to the public. One can have only an impression as to how much good they do, and there is no way of validating that impression. Various attempts have been made to test differences in the person's behavior before and after a course of instruction or reading objectively, so far without verifiable results. Good grades in examinations on the subject matter are of course no valid criteria, nor is subsequent behavior in interpersonal relationships. Obtaining subjective responses from questionnaires as to whether people feel they received benefit from a book or course of instruction provides no real proof. Clearly, what one tries to judge or validate is again of such complexity that fairly thorough individual studies are necessary in order to indicate whether the general trend toward popular education is of any help. Even the term, help, has to be redefined if we break down what we ordinarily call the increment of

knowledge and important factors such as the elevation or diminution of levels of aspiration, changes in anticipation with regard
to marriage, influences toward conformity or nonconformity, increasing or diminishing levels of anxiety, fostering of identifications with a group or with parents, and similar factors that express the phenomena in terms that can be interpreted and
handled.

And so if one wants to talk about the general problem
and ways of educating and of preparing people for marriage,
one must also consider the multiple motives of authors and
teachers which are reflected in their educative policies, often unconsciously. Some of these educative trends are clearly moralis-
tic, others express pressure toward conformity, still others have a
pragmatic trend that seems to promise a person that the more
information he has the happier he will be. It thus seems fair to
say that such educational methods have a value insofar as they
give information. What they might do to help a person overcome
neurotic trends, and thereby equip him better for marriage would
be largely accidental and indirect.

Let us examine next the possible value of what is ordinarily
called "marriage counseling." The field is primarily the domain
of ministers, family physicians, lawyers and others not directly
concerned with psychiatry. For one thing, a goodly number of
people prefer to go to marriage counselors rather than to psychiatrists, and thus avoid the implication of neurosis or need for
treatment. It is also easier for a person to identify himself with a
group of people who have "marital difficulties" rather than an
emotional illness or disturbance. Secondly, the marriage counselor
is by his function more prepared to give advice and suggestions
of an immediate and practical nature, and so to offer a kind of
help that is often met with a comparable state of readiness in the
client to accept it. In some instances that I have had the good
fortune to study, an unexpected factor played a role in this state
of readiness on the client's part and his submissiveness and trust

in turning over the major responsibility to the counselor. The client then unconsciously acted out, according to the counselor's suggestions for improvement of the marital situation, utilizing the advice not so much as a rational plan but as a need to make what the counselor said come true.

In general, what has been said about courses and special books on marital relations may be said of counselors. People in trouble have a certain readiness to follow advice, and especially when this advice consists of information and results in increases in practical knowledge, marital situations have improved. At other times one observes that what happens as a result of advice and counseling is an increase in one neurotic partner's adaption to the other, along with an authorized diminution of the level of aspiration. Indeed this may well be what happens in many instances of marital adjustment that evolve from counseling, psychiatric treatment or the aging process.

The matter of treatment as an aid toward improved marital relations remains to be considered. At the present time this seems to offer the best hope for the future, particularly with our increased psychiatric facilities and increasing skills of therapists in clinics and social work agencies. Nevertheless, even the excellent papers that have been written on psychiatric and psychoanalytic studies of marital difficulties provide little of any definite nature that can be used as a predictive tool. For one thing, the patient who quite willingly and readily goes into a treatment situation with the avowed intent of preserving or improving a marriage has of necessity to take some risk that the treatment might have the opposite result. Even when marital partners go into treatment with the deep conviction that they love each other and want to preserve their marriage, one or the other may come to a different conclusion; although it must be admitted on a clinical basis that in these instances the marital partners have a better chance of staying together.

As clinicians we are undoubtedly impressed with the successes we have had in helping people with their neurotic problems and thus, in turn, helping them with their marital situation. The sequence in the last statement is carefully chosen as the proper one, because logically we look at marital problems as being most likely a part of the process of neurotic interaction, so that a treatment approach should be directed toward helping a person solve his neurosis rather than his marital situation as such. And here the psychoanalytic approach offers most, while at the same time it has the disadvantage of being the most inaccessible of treatments. Hence, we must rely upon psychoanalytically oriented psychotherapy of one sort or another as the medium of choice. Yet, in any treatment, I have been impressed by the unanalyzed factors that continue to play a role even after many neurotic elements have apparently been more or less successfully treated. One of the best examples of this is the person who overcomes neurotic trends sufficiently to obtain a divorce and remarries in an undercurrent attitude of having to prove that the first marriage was a mistake and with the quiet but grim determination to make the second marriage succeed. And it frequently does.

As a final consideration I should like to comment on those accidental social influences that touch upon the psychic structure of individuals and probably have an important influence on marital relations. A realistic attitude is not necessarily a pessimistic one. External influences and realistic life situations do influence certain people in their marital lives. Good fortune and misfortune have both been accidental occurrences that have helped people mature and appreciate each other.

The question may very well be raised, in regard to the last statement, as to whether such episodes as the birth of children, success in work, prestige achieved one way or another, a misfortune which has a binding effect upon a married couple, are elements that make for real maturation, or whether these are simply the results of shifts in symptoms. The question is a valid one only for purposes of investigation and research. The fact is incontrovertible that such external forces have made sufficient changes in the internal economy so as to provide flexibility for adaptation of one neurotic partner to another.

In a similar vein one may remark that improved marriage and divorce laws and other social reforms may increase the possibilities for more felicitous marital relationships. At one extreme, as numerous examples show, the more restrictive and binding the marriage laws, the more will stringent divorce laws reflect an authoritative attitude that makes people bind themselves together in a family unit by command rather than by desire. The aggression that is manifested in divorce would not necessarily be successfully suppressed or diverted into more healthful channels, however. At the other extreme, it is true that experiences with more permissive marriage and divorce laws do show that in themselves they contribute to marital happiness by giving people a greater mobility and control over their own destinies. Such a trend is consistent with a proper setting for the functioning of mature attitudes. Yet the question can arise as to what limits have to be set even for so-called mature individuals. Obviously if social changes are to be effected that would make for a better atmosphere for marital felicity, they will have to be profound measures that take due cognizance of the complexity of interpersonal relationships.

References

1. Goode, William J., *After Divorce*. Glencoe, Illinois, The Free Press, 1956.
2. Korner, A. F., "Theoretical Considerations Concerning the Scope and Limitations of Projective Techniques." *Journal of Abnormal and Social Psychology*, October, 1950.
3. Sapirstein, M., *Paradoxes of Everyday Life,* Chapter I. New York, Random House, 1955.

Psychiatry and the Law in Separation and Divorce

———◆•◆———

RICHARD H. WELS, ESQ.*

THE HEIGHTENED AWARENESS of lawyers, judges and teachers of the interaction of psychiatry and law and the significance of psychiatric diagnosis in legal problems is reflected in a recent announcement by the University of Pennsylvania Law School of the establishment of a program for teaching law students the relationship of psychiatry to modern legal problems. This program, accorded a National Mental Health Act grant, will be under the joint supervision of a law professor and a psychiatrist.

As stated by the Dean: "Lawyers are constantly concerned with behavioral problems. They are called on for advice by the disturbed and the mentally ill, and they see many problems with mental health aspects. They need understanding of human con-

---

* Chairman of Special Committee on Improvement of Family Laws of the Association of the Bar of the City of New York; Member, Interprofessional Commission on Marriage and Divorce Laws and Family Courts.

duct, its motivations and determinants. They need this understanding, not so that they can substitute for psychiatrists, but to be better counselors and judges."

One of the major goals of the project is "to accelerate the reception into law of further advances in psychiatry as they occur in the future by orienting successive generations of law students, many of them destined to become governors, legislators and judges." This statement touches upon a deep-seated problem. For no matter how receptive the individual lawyer or judge may be in his own understanding of the relevance of psychiatric diagnosis to the particular case in hand, he must always work within the framework of the law itself. Unless the law accepts psychiatric diagnosis and acknowledges its relevance, he is powerless to utilize it.

Traditionally, the law is wary of scientific developments and reluctant to accord them legal significance until their accuracy and certainty have been established beyond doubt. For example, years of controversy preceded the acceptance into the law of such scientific fact-finding devices as the lie-detector, and blood tests in paternity cases.

Similarly, the attitude of the law toward sexual deviations will have to be modified on the basis of current psychiatric knowledge. As matters now stand legally, disturbances in sexual behavior arising from faults in the early psychosexual development of the individual are regarded as misdemeanors or felonies. If the law is to remain an instrument of progress and reform, consideration must be given to modern psychiatric diagnosis and treatment in the approach to such cases.

The classic example of the reluctance of the law to accept scientific developments as valid criteria is to be found in the test of criminal responsibility where a court is concerned with a plea of insanity. In 1843 Lord McNaghten, one of the law lords in the House of Lords, laid down the rule that has been consistently followed both in England and in the United States for more than a hundred years, and which may be substantially said to be that the test of criminal responsibility is whether the accused knew the nature and quality of his act and the differences between right and wrong. This has been codified by statute in states such as New York so as to provide that "A person is not excused from criminal liability as an idiot, imbecile, lunatic, or insane person, except upon proof that, at the time of committing the alleged criminal act, he was laboring under such a defect of reason as not to know the nature and quality of the act he was doing; or, not to know that the act was wrong."

Developments in psychiatry in the more than a hundred years since the handing down of Lord McNaghten's rule have of course established that a person well may be insane, and should not be held responsible for his acts, without being able to meet the standards set down by Lord McNaghten. American, as well as English courts, have had to stress repeatedly that legal insanity and medical insanity are entirely different, and have accordingly ruled inadmissible a psychiatric diagnosis which goes beyond the right and wrong test.

The law, however, is slow but not eternally inflexible. Thus, recently in the District of Columbia, the United States Court of Appeals has handed down a decision in which it rejected Lord McNaghten's rule, and held that the defense of insanity could be successfully maintained if the unlawful act was the product of mental disease or mental defect, without limitation to the cognitive faculty alone. Thus, in the District of Columbia, the psychiatric diagnosis may now be received without restriction.

In matrimonial cases, the bulk of which consists of actions for divorce and separation, lawyers are aware of the importance of the psychiatric diagnosis. To cite Howard Hilton Spellman, "In matrimonial cases, the underlying basis of action is emotional. Financial questions, determination of the custody of children, divisions of capital property, and the like are factors flowing from the marital dislocation caused by the original emotional upheaval. Thus, from the very inception of his duties, the lawyer is faced by the need to understand the psychological motivations of the parties." [2]

In no area, however, has the law itself been more rigidly adamant against consideration of psychiatric diagnosis and psychological motivations than in that of matrimonial actions. Our divorce laws today are predicated upon concepts of guilt and punishment. If a spouse has committed an act which the law regards as an offense against the marriage, upon receiving proof of the commission of such an act it punishes the offending spouse by granting the other a divorce.

While a judge in a matrimonial case may want to base his decision on the motivations for acts, instead of on the narrow measure of legal responsibility for proven facts, the law restricts him to exploring the "what" and not the "why."

In most states grounds for divorce include mental cruelty, abandonment, desertion, living apart for a period of years, habitual drunkenness, cruel and inhuman treatment, or even incompatibility, as well as adultery. In New York State the only basis for divorce recognized by the law is adultery. But in all states today, regardless of the precise grounds on which the divorce is sought, the underlying concept and approach are the same.

For example, in a New York divorce action the justice of the Supreme Court (or the Official Referee, a retired judge) before whom the action is being tried has no concern with the real cause of the marital rupture, or whether the marriage might possibly be salvaged. Nor is it within his province to determine whether granting a divorce is in the best interests of the husband, the wife and the children of the particular family unit whose future depends upon his verdict. The fact is that at the present time the law denies him the resources and facilities he would need to exercise such judgment.

It is not even the concern of the court to ascertain the motivation for the adultery—what it was that caused the husband or the wife to seek a sexual relationship outside the home. The sole interest of the court is to determine whether the alleged act of adultery was in fact committed, and whether its commission was either consented to or subsequently forgiven by the other spouse. If a judge finds that such an act of adultery was committed, and was committed without collusion and without condonement by the other spouse, he must grant the divorce. Should he fail to do so, his decision will most certainly be reversed by the appellate courts. It is not difficult for the court to find that such an act of adultery was committed. Under the law, such a finding must be made if it is proven that both the opportunity and the inclination for the commission of such adultery existed.

In states other than New York the approach is precisely the same, though the grounds may be different. In a state where a divorce may be obtained on the grounds that the spouses are living apart, or that one has abandoned the other, or that one is an alcoholic, the court cannot and will not inquire into the motivation for living apart, for abandonment, or for alcoholism. It will restrict itself solely to finding the fact and rendering its verdict accordingly.

Where one party to a marriage is accused of being an alcoholic, the psychiatric diagnosis is of course of particular relevance. Interestingly enough, it is rare that the statutes ever refer to alcoholism or alcoholics as such. In those states where such a condition affords grounds for divorce, the statute always refers to it as "habitual drunkenness," and alcoholics are generally labelled in the statutes as "inebriates" or "drunkards."

Habitual drunkenness is a ground for divorce in all but ten states. The states which do not recognize it as a basis for divorce include such jurisdictions as New York, New Jersey, Pennsylvania, Maryland, the District of Columbia, Texas, Virginia, both Carolinas and Vermont. The definitions and standards utilized by the courts in determining whether habitual drunkenness exists are well established: A person will not be considered an habitual drunkard by reason of an occasional consumption of alcohol, even though it may be excessive. The significant element is proof that the intoxication has become a persistent habit.

The courts have defined an habitual drunkard as one who, by frequent periodic indulgence in liquor to excess, has lost the power or desire to resist an alcoholic opportunity, with the result that intoxication has become habitual rather then occasional. The alleged drunkard need not be continuously intoxicated, nor have more drunken than sober hours, nor become intoxicated every day, nor even at regular intervals. So long as there is a fixed habit and inability to control the appetite, it is unimportant that he may abstain for periods, or is able to attend to his business.

In the jurisdictions where a divorce may be obtained on the ground that the defendant is an habitual drunkard, the law makes a distinction between those cases where the habitual drunkenness came into being after the marriage, and those cases where it existed prior to the marriage. Nowhere may a divorce be granted on such ground if the alcoholic habit existed before marriage.

In determining the existence of alcoholism, evidence as to physical symptoms is considered relevant. Testimony may be given as to the use of violent language and general physical appearance. Such details as the fact that the defendant "looked flushed and blue," that his eyes were blinking, that he had alcohol on his breath, that he staggered in walking are all relevant, as is testimony about his going to a sanitarium for a cure, a change in his personality while drinking and afterward, or unusual acts which he committed while being drunk. Testimony as to conviction for driving while intoxicated, including the results of alcoholic measurement tests used by the police in connection with arrests for such an offense, is also acceptable.

While such testimony is pertinent as to the fact of drinking, it is not sufficient, unless so detailed as to indicate a pattern of

behavior and addiction, to establish the essential fact of persistent habit. Here the testimony of a qualified psychiatrist, with
specific experience in dealing with problems of alcoholism, becomes particularly relevant and determining.

The conventional attitude of the law toward divorce has
been summarized by Judge Paul W. Alexander of the Court of
Common Pleas of Toledo, Ohio, who as Chairman of the Interprofessional Commission on Marriage, Divorce and Family
Courts, and one of the great architects of the law, has advocated
the adoption of an entirely different approach. Judge Alexander
commented:

"When may a divorce be granted? Only when one party is
guilty. That is the only criterion fixed by the law in every state.
Guilt is made the cause of divorce. What form of guilt is specified by the 'grounds,' the overt acts or commissions which the
law says shall be sufficient cause—things we used to call sin.
True, specific sinning is often the *proximate* cause, the last straw,
but rarely, if ever, the ultimate cause. When mamma nags or is
cold and drives papa to the corner 'nite club' and a 'drug-store
blonde,' the blonde is only a superficial, intervening cause (in
more ways than one). She is really effect, not cause, of the
ostensible marriage failure. And it isn't necessarily a marriage
failure—yet—both parties to the contrary notwithstanding. And
divorce is not necessarily the right prescription, no matter how
loudly either or both parties may proclaim it is." [1]

Both the Interprofessional Commission and the Special
Committee on Marriage and Divorce and Family Courts of the
American Bar Association have urged a change in divorce laws,
and the idea is gaining wide acceptance among bar associations,
judges, and the public, although it is not yet reflected in the
laws of any state. This new approach is similar to that taken in
the children's courts in cases of juvenile delinquency.

For at least two generations, the law has abandoned the
guilt and punishment dictum in dealing with children generally
and juvenile delinquents in particular. It has substituted instead

the concept of diagnosis and therapy. In the juvenile court the law recognizes that the function of the court is not merely to ascertain the guilt of the child and administer punishment. It has accepted the thesis that the function of the court is to determine what caused the child to commit its antisocial and unlawful act, and then to administer the therapy that will bring about the rehabilitation of the child. In this regard Judge Alexander states:

"The American Bar Association Committee proposes to transform the divorce court from a morgue into a hospital; to handle our ailing marriages and delinquent spouses much as we handle our delinquent children—for often their behavior is not unlike that of a delinquent child and for much the same reasons. Instead of looking only at the guilt of the defendant, it proposes to examine the whole marriage, endeavor to discover the basic causative factors, and seek to remove or rectify them, enlisting the aid of other sciences and disciplines and of all available community resources.

"The fresh approach, the new philosophy would be signified by the very titling of the case; instead of it being *John Doe v. Mary Doe,* it would be titled *In the Interest of the John Doe Family.* There would be no plaintiff, no defendant—only an applicant or joint applicants. The application would not be for divorce but for the remedial services of the state. Petition for divorce would be permitted only after complete investigation and report. The new plan would take over almost bodily the entire philosophy, procedure, and techniques of the juvenile court. As in the juvenile court the criterion (fixed by law as well as philosophy) is 'What is best for this child?' so in the divorce court the criterion would be 'What is best for this family?' . . .

"Like the juvenile court the family court would require an adequate staff of trained technicians and skilled specialists, such as the social case-worker, psychiatric case-worker, clinical psychologist, psychiatrist, marriage counsellor, and others, and, of course, a proper clerical force. Most important, the court would require a specialist judge or judges. No court can be expected to rise above its judge. No matter how able a lawyer he may be or how filled with the spirit of altruism, he will have to school himself in quite a number of fields of learning and disciplines for which his legal training and experiences have not prepared him. Among these are social casework, group work, counselling, diagnosis and therapy, several branches of psychology (especially so-called abnormal psychology), penology, criminology, the basic principles of psychiatry, medical casework, community organization, child and family welfare, and some others."

Obviously, in this broad new plan advocated by the American Bar Association and by the Interprofessional Committee, psychiatric diagnosis is a major and key factor. Although its basic concepts may confidently be expected to be absorbed into our laws and statutes within a generation, it does not today represent the law of a single state. Nevertheless a beginning has been made. A recent editorial in the New York *Times* hails the opening of the new Family Part in the New York County Supreme Court with Justice Samuel M. Gold presiding. All cases involving family relations will go before Judge Gold whose primary aim will be reconciliation of the parties involved. In this task he will be assisted by trained social workers and psychiatrists. A special committee of the New York City Bar Association will be available for consultation and will help in providing the aid of specialists in social health.

Actions for separation, like actions for divorce, are brought by one spouse against another because of some offense committed against the marriage by the defendant spouse. They, too, are presently geared to the concepts of guilt and punishment. In many states the same acts which constitute grounds for divorce are also grounds for a separation, while other acts, not considered sufficient cause for the granting of a divorce, are also made grounds for separation.

For example, in New York the adultery which is grounds for a divorce is also grounds for a separation. In addition, actions for separation may be upheld because of the cruel and inhuman treatment of one spouse by the other, or conduct on the part of one spouse making it unsafe and improper for the husband and wife to cohabit, or abandonment and failure on the part of a husband to provide for his wife.

An action for separation differs from an action for divorce in that, although the court may decree that the parties shall live separate and apart, the marriage itself remains undissolved, and neither husband nor wife may remarry.

As in divorce cases, in an action for a separation the court may only find the facts at issue, and may not inquire into the motivation behind such acts, except to the extent that there may be some adequate justification or provocation for them by the behavior of the other spouse. Nevertheless, in proving the existence of some of these basic facts, the psychiatric diagnosis is relevant.

The New York statute, as has been noted, permits a court to decree a separation where one spouse has been guilty of cruel and inhuman treatment, or where such spouse's conduct has been of such a character that it becomes unsafe for the other to cohabit with the offending spouse. In determining the existence of such facts there may be some relevance in the psychiatric diagnosis. Of course, where the conduct complained of is physical in character, the court does not require a psychiatric diagnosis to evaluate the proof. But the New York courts have held that the right to a separation on grounds of cruel and inhuman treatment is not confined to actual violence or threats of violence, that it may include mental pain, and that certain conduct may produce such mental pain as to be even more cruel and inhuman than the infliction of physical pain. However, it must be shown that the conduct was of such nature as to seriously affect the spouse's health.

Obviously, this requirement of the law that such conduct must affect the health of the plaintiff or render it unsafe for the husband and wife tó cohabit invites psychiatric testimony. Although the New York courts do not go as far as those in Pennsylvania and Minnesota, where the courts have held that medical testimony is essential for the establishment of such facts, the testimony of a qualified and competent psychiatrist is, of course, relevant and desirable.

It should also be noted that in addition to actions for divorce and separation the New York law provides that a marriage may be dissolved where the husband or wife has been incurably insane for a period of five years or more. Here the law itself specifically recognizes the relevance of the psychiatric diagnosis.

Section 7 of the Domestic Relations Law states that no judgment may be given dissolving such a marriage "until, in addition to any other proofs in the case, a thorough examination of the alleged insane party shall have been made by three physicians who are recognized authorities on mental disease, to be appointed by the court, all of whom shall have agreed that such party is incurably insane and shall have so reported to the court."

Another form of matrimonial litigation is an action for an annulment. Where obtained, such action results in the dissolution of a marriage and in the consequent freedom of the parties to remarry. Unlike an action for a divorce, which is predicated upon the conduct of a spouse during the life of the marriage, an annulment action is concerned with those facts and conditions existing at the time of the inception of the marriage. Therefore an annulment may be awarded only where one party entered into the marriage because of the other's fraudulent and false representations of a major character, or where, for one reason or another, it turned out that one of the two spouses was incapable of consummating the marriage.

Interestingly enough, statistics show that three-fourths of the annulments in the United States are obtained in New York and California, and that in New York there is a fairly constant ratio of one annulment action for every two divorce actions. The popularity of California annulments has been explained by the fact that while there may be no remarriage for a period of a year after a divorce in California, there may be immediate remarriage after an annulment in that state. In New York, the explanation seems to lie in the fact that adultery is the only ground for divorce.

Impotency is specifically recognized by New York law as an incapacity that precludes entering the marital state. Since the courts have found that impotence may exist because of psychological factors or nervous tension, the psychiatric diagnosis plays a substantial role in such cases. Similarly, where it is alleged that a spouse was insane at the time the marriage was entered into, the psychiatric diagnosis is also of considerable significance and is the controlling factor.

In all matrimonial actions where there are children of the marriage who are minors, the court has the power and responsibility of making whatever decision as to their custody it deems to be in their best interests. The paramount factor in such a decision is, of course, the welfare of the children, with the interests of either or both parents necessarily subordinated to those of the children. In determining what is best for them, the court must frequently deal with the subsidiary question of the fitness of one or both parents to take custody.

Courts are particularly sensitive to problems relating to the custody of children, and approach such matters with careful consideration and thoroughness. Judges who are called upon to make decisions in custody cases are invariably anxious to do the best they can for the children, realizing that they are the principal victims of broken homes. Fortunately, when the question of custody comes up in the Children's Courts under neglected children statutes, those courts are equipped with adequate psychiatric and investigative resources which enable the judge to make an informed judgment.

In the highest courts which have jurisdiction of divorce and separation actions, such resources are completely lacking. Sometimes a judge will call upon other courts—the Children's Court, the Domestic Relations Court, or even the probation departments of the criminal courts—for the use of such facilities and personnel to assist him in making his decision. Lacking its own facilities, the court will also welcome the assistance of such expert psychiatric testimony as the parties themselves may be able to make available. Such testimony may be concerned with the specific neuroses or psychoses or other behavioral pattern of the particular parent. The focal point, however, is the impact of a given condition and behavioral pattern upon the children.

The testimony of a psychiatrist is, of course, subject to the statutory provisions which make the testimony of a physician privileged. Such privilege may not be waived or released without the consent of the patient. It should be noted, however, that since the law creates such a privilege only in the cases of physicians, lawyers and priests, it is not available to a psychologist or psychoanalyst who is not a physican.

In describing the duties of a modern judge, Judge Francis L. Valente of the New York Court of General Sessions recently wrote: "There is a tremendous body of new knowledge that must be considered. The enlightened judge, sensitive of his obligations, must familiarize himself with the advances in science, in psychology, for instance, so that he can assay a man's personal liabilities and assets—learn what motivated the criminal act—study him in the light of his personal and social history, his family life, his weaknesses, his emotional stability, his aspirations even. . . ."

Yet the fact remains that even the most sensitive judge, no matter what his anxiety and concern for the individuals and families that come before him, must inevitably bow to the law. If the law attaches no value to the psychiatric diagnosis and does not permit it to be a factor which he may consider in making his determination, he must exclude it from his decision. As we have seen, except in the subsidiary question of custody, the law for the most part imparts no significance to the psychiatric diagnosis in actions for divorce and separation.

What has been discussed so far relates only to court actions.

It may surprise some readers to learn that in most states, including New York, more than ninety-five per cent of all matrimonial actions are uncontested. This means that in the majority
of these actions for divorce, separation and annulment the husband and wife have resolved their differences between themselves before coming to court. They have arrived at a private
agreement as to whether they shall obtain a divorce, where they
shall obtain it, what provisions shall be made for the support
and maintenance of the wife and children, as well as the custody
of the children. Their purpose in coming to court is only to
give some legal implementations and recognition to decisions
that have already been made.

Most matrimonial cases, therefore, present no issues to the
court or to the judge. A minimum of testimony is offered, just
enough to meet the requirements of law. Witnesses are not subjected to cross-examination, and the judge does not have to sift
the credible elements from conflicting testimony. Separation
agreements embodying the private arrangements of the parties
as to support and custody are submitted to the 'court, and, in
the absence of controversy, are almost invariably approved and
incorporated in the court's decree. No psychiatric diagnosis is
ever presented in these cases which are tantamount of divorce
by consent.

Yet it is precisely in these uncontested actions that a psychiatric diagnosis might play a constructive role in formulating a
decision as to whether or not to grant a divorce, which parent
should have custody of the children, and whether it is advisable
for the husband and wife to make another effort at reconciliation.

At some point from the time of the disturbance in the marriage to the moment the family appears in the courthouse, the
husband and wife have gone to the offices of their respective
lawyers. In some instances it is obvious to a lawyer that a situation exists which plainly calls for judicial recognition of the
end of a marriage long since dissolved. Here there is no possibility

of salvaging the family, correcting or eliminating those factors which have created the deterioration of the marriage, or effecting a reconciliation. In other cases the lawyer seeks the aid of psychiatrists or members of other allied professions. Calling in a psychiatrist or psychologist will often help the lawyer to find an ultimate solution that may do away with the necessity for bringing the case to trial.

The lawyer performs no more difficult task than when he is called into a matrimonial situation. It presents him with his greatest challenge. If he is consulted before the rupture is complete and the break past healing, he will marshall all available resources and facilities in an effort to keep the family together. He will form a team, with his adversary as co-captain, where possible, with psychiatrists and other skilled specialists in an effort to accomplish the best possible result for the family as a whole and to achieve the maximum happiness and stability possible for that family.

Even where there is no chance of keeping the family together, the lawyer will realize the importance of obtaining proper psychiatric assistance for his client so that there may be some adequate understanding of the factors that brought about the bankruptcy of the marriage. Only by so doing can the materials be provided with which to found a new, sound and happy life, and insure the children of the marriage adequate provision for their emotional, psychological and material needs, with a minimum of scar and injury.

Such instances are not reported in the law books either because they were uncontested when they came to the courthouse, or, if contested, because the law did not consider psychiatric diagnosis relevant. But there are in lawyers' files records of thousands of cases for divorce and separation in which psychiatric diagnosis, though not accepted by the law, was a highly important factor.

Most lawyers prefer to effect a settlement of property rights and support provisions through a separation agreement, rather than through a court action for a separation or a divorce. Eventually both methods achieve the result of permitting the parties to live separate and apart. But the separation agreement, with its subsequent uncontested court action, avoids the inevitable antagonisms that come from laying bare marital wrongs and hurts in formal affidavits and in the courtroom. For the same reason, separation agreements do not preclude the possibility of the parties coming together again. On the other hand, once a contested matrimonial action is under way, the chances for reconciliation, as all lawyers know, are exceedingly slim.

The law often provides that a divorce decree shall not become final for a period of three months to a year after the initial court decision has been made. This is done in order to permit the parties to review the wisdom of their decision and perhaps effect a reconciliation. Although the law and the courts deliberately leave the door open, the privilege of reconciliation is taken advantage of rather infrequently and then almost always by those who have been in an uncontested rather than a contested action.

In this writer's opinion, most lawyers would prefer to see matrimonial matters, as well as all other matters pertaining to the family, confined to courts in the state where the family actually lives. Further, these courts should be equipped with adequate personnel and facilities to make sure that all of the facts are accurately, fully and fairly presented. The court should be convinced that there is no possibility of reuniting the family, and that a divorce is in the best interests of the entire family, including the children, before it awards a decree of divorce.

There can be no doubt that the personnel and facilities required by such a court would necessarily include a competent psychiatric staff. Experience has shown that an expert in psychiatric matters can make a notable and constructive contribution to the solution of family difficulties. Such measures would most certainly decrease the divorce rate and foster the stability of marriage and the American home.

References

1. Alexander, Judge Paul W., *American Bar Association Journal,* February, 1950.
2. Spellman, Howard Hilton, *Successful Management of Matrimonial Cases.* New York, Prentice-Hall, Inc., 1954.

Index

ambivalence (*cont.*)
  toward parents, 53, 106, 218, 226,
    275
American Bar Association, Special
  Committee on Marriage and
  Divorce and Family Courts,
  334-36
American Psychoanalytic Association,
  39
anaclitic object choice, 58
analyst, *see* psychoanalyst
annulment, 336-37
Antabuse, 153, 162, 164, 166
"anxious family," 85-87

Bacon, Seldon, 149, 151-52, 167, 168;
  see also Straus, R.
Baker, S.M., 159, 167
Balint, Michael, 167
"basic anxiety" (Horney), 155
Beatman, Frances Levinson, 262-68
Bellevue Intelligence Test, *see* Verbal
  Wechsler-Bellevue test
Bergler, Edmund, 104, 124
Berkowitz, Sidney J., 261
Bird, H.W., *see* Martin, P.A., and
  Bird, H.W.
Bleuler, Eugen, 125, 143
body-phallus fantasy, *see* wife-penis
  fantasy
Boggs, M., 159-60, 167
"brief contact," 287
Burlingham, Dorothy T., 100, 224,
  234
Bychowski, Gustav, 135-47
  and Despert, J. Louise, 289
Burrow, Trigant, 304, 310

casework and caseworker:
  application for, 32-33
  attitude toward, 279-80
  -client relationship, 266
  counseling, *see* casework treatment
  diagnosis, 235-61
  follow-up of, 284, 287
  functions, 263
  psychoanalytic concepts in, 235-43
  research in, 286-88
  supervision in, 236, 286
  technical tasks of, 247-49

casework and caseworker (*cont.*)
  treatment, 124, 236-37, 262-89,
    296
castration anxieties, 68, 70, 72, 106,
  115, 120, 122
Children's Courts, 337
child and children:
  buffer role of, 50-51
  as confidant, 53
  custody of, 339-40
  effect of alcoholism on, 148, 150,
    156-57, 162
  effect of divorce on, 45, 54, 55,
    312, 341
  effect of marital discord on, 13, 33,
    34, 42, 44-56, 132, 227, 229,
    232-33, 270, 339
  legal attitude toward, 329, 334,
    337-38
choice of mate, *see* marital choice
"choice of the neurosis," 64
chromatic color responses (Rorschach
  test), 204-206
Clinical Sector Psychotherapy, 296
clinics for alcoholics, 154
coitus interruptus, 112
communication between marital partners, 6, 224, 249
Community Service Society, Division
  of Family Services, 235-36,
  242-43, 244
compatibility, definition of, 192-93
compensatory mechanisms, 231
competitive marriages, 29-30, 184
complementary personality patterns,
  between married partners, 81-
  85, 127-28, 272-85
  aggressiveness-dependency, 81
  dependency-support, 84, 98, 127-
    128
  detachment-love demand, 82-83,
    98
  domination-submissiveness, 83, 98
  helplessness-considerateness, 83-
    84
  neurotic, 98-100
  satisfaction from, 84
  in parent-child relations, 85-88
  in sibling relations, 85-88
compulsive drinking, *see* alcoholism

St. Scholastica Library
Duluth, Minnesota 55811